APOLLO 1

APOLLO 1

THE TRAGEDY THAT
PUT US ON THE MOON

RYAN S. WALTERS

REGNERY
HISTORY
Washington, D.C.

Regnery History™ is a trademark of Salem Communications Holding Corporation
Regnery® is a registered trademark of Salem Communications Holding Corporation

ISBN: 978-1-68451-094-8
eISBN: 978-1-68451-147-1

Library of Congress Control Number: 2021953048

Published in the United States by
Regnery History
An imprint of Regnery Publishing
A division of Salem Media Group
Washington, D.C.
www.RegneryHistory.com

Manufactured in the United States of America

10 9 8 7 6 5 4 3 2 1

Books are available in quantity for promotional or premium use. For information on discounts and terms, please visit our website: www.RegneryHistory.com.

To the families of Gus Grissom, Ed White, and Roger Chaffee:
You made the ultimate sacrifice to put America on the moon. A
grateful nation will forever be in your debt.

"Ad astra per aspera."
A rough road leads to the stars.

"There's always a possibility that you can have a catastrophic failure, of course; this can happen on any flight; it can happen on the last one as well as the first one. So, you just plan as best you can to take care of all these eventualities, and you get a well-trained crew, and you go fly."
—Gus Grissom

"People might look at our work as being perhaps dangerous, or risky of sorts, but I think we train in it and work in it so much and understand it well enough that we don't look at it from this viewpoint. We accept the risks."
—Ed White

"There's a lot of unknowns and a lot of problems that could develop or might develop and they'll have to be solved. And that's what we're there for. This is our business to find out if this thing will work for us."
—Roger Chaffee

CONTENTS

MAJOR PLAYERS

Clinton Anderson—U.S. senator from New Mexico; chaired Senate Space Committee hearings

Bobby Baker—Powerful protégé of Lyndon Johnson and secretary for the Senate majority; known as the "101st Senator"

Fred Black—Washington lobbyist for North American Aviation

Frank Borman—Astronaut, Gemini 7; member of Apollo 204 Review Board; commanded Apollo 8, the first flight to the moon

Martha Chaffee—Wife of Roger Chaffee

Roger Chaffee—Astronaut, Apollo 1 pilot

Walter Cunningham—Astronaut, Apollo 1 backup crew, Apollo 7 pilot

Kurt Debus—Head of the Kennedy Space Center at Cape Canaveral

Donn Eisele—Astronaut, Apollo 1 backup crew; originally on the Apollo 1 prime crew; senior pilot on Apollo 7

Robert Gilruth—Head of the Space Task Group; director of the Manned Spacecraft Center in Houston

Betty Grissom—Wife of Gus Grissom

Gus Grissom—Astronaut, Apollo 1 commander; Mercury 4 and Gemini 3

Lyndon B. Johnson—36th president of the United States; in office at the time of the fire

John F. Kennedy—35th president of the United States; assassinated on November 22, 1963

Robert Kerr—U.S. senator from Oklahoma and millionaire oil man; helped North American Aviation receive the contract to build the Apollo spacecraft

Chris Kraft—NASA flight director; director of flight operations in Houston

Walter Mondale—U.S. senator from Minnesota; criticized NASA during the Senate hearings in 1967

Lola Morrow—Secretary to the astronauts

George Mueller—Associate administrator for Manned Space Flight

Rocco Petrone—Director of Launch Operations at the Kennedy Space Center

Sam Phillips—Air Force general; head of the Apollo Program in Washington, D.C.

Wally Schirra—Astronaut, Apollo 1 backup crew; Mercury 8, Gemini 6, Apollo 7

Robert Seamans—Deputy administrator of NASA

Joe Shea—Head of the Apollo Spacecraft Program in Houston

Deke Slayton—Original Mercury astronaut; director of Flight Crew Operations in Houston

Tom Stafford—Astronaut, Apollo 1 support crew; Gemini 6, Gemini 9A, Apollo 10, Apollo-Soyuz Test Project

Harrison "Stormy" Storms—Headed the Space Division at North American Aviation

Olin Teague—Texas congressman; chaired House Space Committee hearings

James Webb—Administrator of NASA; appointed by Kennedy, served under Johnson until 1968

Ed White—Astronaut, Apollo 1 senior pilot; Gemini 4 pilot, the first American to walk in space

Pat White—Wife of Ed White

John Young—Astronaut, Apollo 1 support crew; pilot on Gemini 3 with Gus Grissom; command module pilot on Apollo 10 and commander of Apollo 16, the ninth person to walk on the moon

INTRODUCTION

O n January 27, 1967, astronauts Gus Grissom, Ed White, and Roger Chaffee climbed into a new spacecraft perched atop a large Saturn rocket at the Kennedy Space Center. They were in Florida for a routine dress rehearsal of their upcoming launch into orbit, then less than a month away. Their mission, set for February, would inaugurate the new Apollo program.

All three astronauts were experienced pilots with dreams of one day walking on the moon. Little did they know that once they entered the spacecraft that cold winter day, they would never leave it alive. The mission meant to herald the dawn of the Apollo program would lead it to near failure before it ever got off the ground.

Project Apollo had one goal: land a man on the moon and return him safely to the earth—something mankind had often dreamed of but had never achieved. The moon has captured man's imagination for millennia as an object of fascination, wonder, and scientific study. To some it is a deity to be worshipped; to others, the inspiration for a lover's serenade; and for a chosen few in the 1960s, it was a military objective in a worldwide battle for global supremacy.

A century before the great moon race, the moon captured the full attention of French novelist Jules Verne, a man with a vision that was truly ahead of his time. "The moon, by her comparative proximity, and the constantly varying appearances produced by her several phases, has always occupied a considerable share of the attention of the inhabitants of the earth," he wrote in his epic novel *From the Earth to the Moon*.[1]

The year was 1865. While America was ending four years of bloody conflict, Verne was dreaming of men from earth peacefully traveling to the moon, nearly a century before such a journey became the center of the Cold War world. Verne prophetically envisioned that Americans, not Europeans, would achieve such an astonishing feat. "As for the Yankees, they had no other ambition than to take possession of this new continent of the sky, and plant upon the summit of its highest elevation the star-spangled banner of the United States of America," he wrote. They would do so, he imagined, using a cannonball-type of spacecraft, fired from an enormous space gun named *Columbiad*, which would leave Florida, make it to the moon in four days, and return to the earth by splashing down in the Pacific Ocean.

A writer but also a man of science, Verne had discussed his ideas with French scientists and had worked out calculations for a lunar flight, which, along with much of his adventurous vision, turned out to be surprisingly accurate. The Apollo 11 spaceship would be named *Columbia*, and the trip to the moon during Apollo would take three days. Verne accurately predicted the speed required to break out of earth's gravitational pull and fly to the moon, calculating a necessary speed of 12,000 yards per second, almost identical to the 35,500 feet per second at which Apollo's translunar injection burn (TLI) would eventually push the spacecraft.[2]

Although viewed by many at the time as a muser of pure fiction and unattainable fantasy, Verne nevertheless possessed tremendous optimism about mankind's potential. "In spite of the opinions of certain narrow-minded people," Verne wrote, "who would shut up the human race upon this globe, as within some magic circle which it must never outstep, we shall one day travel to the moon, the planets, and the stars, with the same facility, rapidity, and certainty as we now make the voyage from Liverpool to New York!" But the perils of early space flight were far more serious than Verne could have imagined, as the world would learn in the twentieth century.[3]

As technology advanced over the course of the ensuing decades, mankind curiously looked up at the moon and wondered if it might be possible to make the round-trip flight of half a million miles, just as Verne had predicted. Verne's work played no small part in planting this great hope in the bosom of ambitious figures. German rocket engineer Werner von Braun, who came to America after World War II and helped accelerate the U.S. space program, was enamored with Verne and his writings. During the 1950s, while U.S. space efforts lagged, von Braun, eager to push his ideas forward, teamed up with a visionary of a different sort who shared his dream of space exploration: Walt Disney.

Like von Braun, and Verne before them, Disney thought our species might conquer spaceflight one day. He befriended von Braun, who served as a technical consultant to Disney and appeared on three of Disney's television shows—"Man in Space," "Man and the Moon," and "Mars and Beyond." These programs, airing in 1955, showcased the vast research von Braun had been conducting in regard to the future of spaceflight. Millions watched the broadcasts and were awed by what they saw—ideas that seemed far-fetched being advocated by men who believed they were achievable, and soon. When completing his first theme park, Disneyland in California, Disney announced plans for "Tomorrowland," a second theme park packed with new experiences developed with von Braun's assistance. "Our 'Flight to the Moon' takes place sometime in the future, when travel to outer space will be an everyday adventure," he said on the broadcast announcing the new park.[4]

By the start of the 1960s, projects that had started as dreams were becoming reality. The world's two most powerful nations, the United States and the Soviet Union, were taking their first steps into space, while laying out elementary plans for lunar flight in a budding space race competition. Each party hoped that its success would showcase which superpower would ultimately emerge as the global hegemon.

The United States used good ole fashioned American ingenuity, organization, and ambition to run the space race, crafting a methodically planned program to compete with the Russians. NASA was created in 1958 to organize the step-by-step effort. The first step was getting a man into space, an effort NASA named Project Mercury. Project Gemini would soon follow to perfect the techniques needed to reach the moon, while Project Apollo had the sole objective of successfully completing a lunar landing.

Despite the well laid-out plan, the United States stumbled out of the starting blocks, looking doomed to be the sure loser of the space race. But, because of their methodical approach, the Americans eventually surpassed the Soviets in the race for the moon. By the end of 1966, NASA had successfully completed its first two manned programs: Mercury and Gemini. More than two dozen American astronauts had flown 16 missions, spent 1900 hours in earth orbit, traveled millions of miles, remained in space for long durations to test man's potential endurance in zero gravity, conducted rendezvous and docking maneuvers, and perfected spacewalking. Despite the loss of three astronauts in plane crashes—Ted Freeman in 1964, who had not yet been assigned to a mission, and Elliot See and Charlie Bassett, who were killed in 1966, several months before their scheduled Gemini 9 flight—no space fliers had perished on an actual mission or even in training for a spaceflight, and no American astronaut had ever died inside a spacecraft despite a few close calls.

Confidence was growing that perhaps routine lunar and interstellar travel might soon become reality. NASA was beginning to broaden the dream of a modest lunar landing to include the development of orbiting telescopes, permanent space stations, moon bases for long term habitation, and trips to Mars and beyond. The original plan for Apollo was ambitious, with more than two dozen manned missions. The heavens were hardly the limit for America's space program.

With the successful completion of Mercury and Gemini, the final leg of the race to the moon was set to begin on February 21, 1967, with the flight of what NASA billed as AS-204, unofficially known as Apollo 1. As America prepared to meet President John F. Kennedy's pledge of getting to the moon and back by the end of 1969, the public received horrible news just three weeks before the first launch of the Apollo program: the three-man crew of Apollo 1, Gus Grissom, Ed White, and Roger Chaffee, had died in a tragic flash fire inside the new Apollo spacecraft during a routine simulated-launch test.

NASA's goal of getting to the moon, and the dream of an ambitious spaceflight program, suddenly looked light-years away. The news shook the entire country as well as the nation's resolve to pioneer spaceflight, and millions of Americans mourned the loss of three national heroes.

Apollo 1 commander Virgil Ivan "Gus" Grissom was one of America's original astronauts, chosen to participate in Project Mercury after years as an Air Force test pilot. He flew the second Mercury mission, a sub-orbital flight on July 21, 1961, losing the spacecraft when the hatch prematurely blew and sank to the bottom of the Atlantic Ocean. Questioned and criticized, an embarrassed Gus bounced back and commanded the first Gemini mission in 1965, and was asked to lead the first Apollo flight set for early 1967. If all went well, Gus had a private assurance that he would be the first man on the moon.

Senior pilot Edward Higgins White II joined the NASA astronaut corps in 1962, as a member of the Second Group, the "New Nine," along with notable astronauts Neil Armstrong, Jim Lovell, Tom Stafford, Pete Conrad, and Frank Borman. A graduate of West Point and the U.S. Air Force Test Pilot School, Ed White had logged thousands of flight hours before joining the space agency. In 1965 he flew as a pilot aboard Gemini 4 and became the first American to walk in space. Rather than command his own Gemini mission, which all other Second Group pilots did, Ed was chosen by Gus to serve on the Apollo 1 crew.

Pilot Roger Bruce Chaffee was the mission's rookie, having joined the corps in 1963 as a member of NASA Astronaut Group Three. Although he did not fly during Project Gemini, Roger served as one of the capsule communicators, known as CapCom, during White's Gemini 4 mission. Before becoming an astronaut, Roger flew numerous operations as a Navy pilot over Cuba from 1960 to 1962 when tensions with the island nation were at their peak.

Although the fire and loss of three brave Americans was truly tragic for the nation's space program and set it back nearly two years, the tragedy, and the subsequent investigations, did have a silver lining: the Apollo spacecraft was completely overhauled—including the hatch which now could be opened in seconds—resulting in a stellar flying machine to take men to the moon. Experts widely agree that without the fire and the ensuing re-design of the spacecraft, America never would have made it to the moon before the end of 1969, Kennedy's deadline. And, with growing public opposition to the Apollo program as a whole, the United States may have failed to reach the moon at all.

My main objective with this book is to show exactly how important to the eventual success of the Apollo program the changes made after Apollo 1 were, while uncovering the truth in a positive portrayal of the astronauts, the disaster that killed them, and its aftermath. This is a work of real history, a book that aims to provide a bevy of information with a balanced hand. I love NASA, the space program, and the astronaut heroes who risked their lives getting to the moon, but I do not shy away from doling out criticism where it is warranted, including NASA.

Even though the literature on space history is quite vast, and is growing every year, there has only been one major book-length work focused exclusively on the Apollo 1 tragedy, *Murder on Pad 34,* by Erik Bergaust, published in 1968. It received heavy criticism from NASA, which noted that it was "largely a journalistic rehash of criticism...coming from Congress and the media, with very little new commentary or analysis and no new factual information," which concluded that "the human and fiscal sacrifices made in Project Apollo have been in vain, since the Soviet Union (seen as the reason for Apollo) may not be going to the Moon at all." The provocative title seemed to be a way for the publisher to gain press coverage and sell copies, or at least that is what has been alleged in the decades since the book's publication. But Bergaust himself was certainly an anti-Apollo partisan. Consider how he closed out the book: "So the race is on. It never really came to a halt. The tragic Apollo accident hasn't actually changed anything. The crash program as the admitted cost of possible future fatalities, future murders on future Pad 34's, is being pushed as hard as ever." But, as we will see, things did change after Apollo 1, and for the better.[5]

A second work that touched on Apollo 1 was published the following year, in 1969, by Edmund H. Harvey and Erlend A. Kennan—*Mission to the Moon: A Critical Examination of NASA and the Space Program.* The subtitle gives away its obvious bias. NASA rightly opined that the book "does not provide a balanced account of the lunar landing program or NASA. Instead it is filled with critical asides" and is "long on hyperbole and short on reasoned analysis." The authors questioned NASA's "technical and managerial ability" and accused the space agency of an uncaring attitude toward its own astronauts. "In spite of countless assertions to the contrary, NASA simply did not take 'every step humanly possible to maintain the safety' of the astronauts, either in space or on the ground," they wrote. They also alleged that NASA continued to maintain "the same old pre-fire mentality," which is a common theme with these diatribes.[6]

Another book published in the 1960's also took a negative view of the events in question and the space program in general—*Journey to Tranquility: The History of Man's Assault on the Moon* by British writers Hugo Young, Bryan Silcock, and Peter Dunn. Although it is not solely about Apollo 1, large portions of the book are devoted to the tragedy, and it is essentially "a ponderous 'anti-Apollo' broadside" that "seeks to cast aspersions on the entire space program," using "the Apollo fire that killed three astronauts as the evidence that 'proves' the dishonesty and criminal behavior of NASA and other space advocates." The anti-American bias of the book's foreign authors leaps off the page from the very beginning.[7]

Although each of these books, no matter how biased or slanted they may be, contains some useful information and draws interesting conclusions, Americans who take pride in the nation's space program would be disappointed in the overall narrative frame the works on Apollo 1 take. They are polemics more than anything else, written in the heat of a political debate that has since been resolved. As NASA historians rightfully concluded, the overall motive of all three of these accounts was to incite public opinion against NASA and the entire space program. The authors tried to make the case that NASA had learned nothing from the accident. But how could they have known in 1968 and 1969 what NASA had learned? The eventual success of the Apollo program proves the contrary: NASA learned a great deal from its early failure and incorporated important criticism into their operation.

For shorter versions of the tragedy, there are a number of other books, including Charles Murray and Catherine Bly Cox's *Apollo*, and Andrew Chaikin's seminal work on the Apollo program, *A Man on the Moon: The Voyages of the Apollo Astronauts*, which devote significant attention to the fire. Most astronaut memoirs and oral histories also cover the fire. These books do not take so harsh a view of the tragedy or NASA and the moon program.

As for the film industry, the 1998 HBO series, *From the Earth to the Moon*, based on Chaikin's book and produced by Tom Hanks and Ron Howard, who also teamed up three years earlier to film *Apollo 13*, devotes an entire episode to Apollo 1, as does ABC's series *Astronaut Wives Club*. The tragedy was also featured at the opening of *Apollo 13* and in the 2018 film *First Man*, the biopic of Neil Armstrong.

"Nothing is new, it is just forgotten," the old adage goes. Some may contend that given the previous literary work there is nothing new to add to the story of the Apollo 1 fire and the subsequent investigations and their findings, as well as the political intrigue swirling around the awarding of the prime contract for the spacecraft, but since the publication of those initial critical examinations, fifty years of new material unavailable to earlier writers has been released—numerous memoirs by astronauts, NASA officials, and members of the press, and an abundance of oral histories by NASA and other academic venues. I was also able to speak to a few persons with first-hand knowledge of individuals and events. These valuable materials represent a gold mine of new information and opinions on the tragedy. I draw heavily from these primary sources, most particularly on the words of the astronauts in the early space program and those closest to them, to construct my own narrative in my own way, to tell the story with fresh sources aimed at a new generation of Americans, most of whom know very little of the tragedy.

Using this new material as well as information previously published, this book will focus on the tragedy of Apollo 1 and the lives of each crew member, and also on the history of the space program as a whole and how Apollo 1 impacted it. In other words, this is a book about America's pioneering space program through the prism of Apollo 1 and its crew.

A work of this magnitude, utilizing new material and fresh perspectives from major participants, is very important for the history of the space program. My aim is to gather all knowledge and information under one literary roof for the sake of posterity. It is a work of popular history, primarily geared toward the general public, and I labored long and hard to produce a work of history as free of technical jargon and legislative mumbo-jumbo as possible.

The deaths of Gus Grissom, Ed White, and Roger Chaffee should never be forgotten. Instead, they should be honored for their noble lives, service to their country, and sacrifice for the good of all mankind, for in a very real sense, the tragedy that took their lives paved the way for America's future success in space. This is their story.

THE FOUNDATION:
EARLY RACE FOR SPACE

With the end of World War II, America was sitting atop the world as its sole superpower. The United States boasted the world's greatest economic engine, a massive "arsenal of democracy" that had almost single-handedly defeated Nazi Germany and Imperial Japan. It could also brag about its lethal military machine backed by industrial might and the newest form of utter destruction: the atomic bomb. Within a quarter of a century, two global conflicts, claiming as many as a hundred million lives, reeking incalculable destruction across Europe and Asia, had devastated the once great powers of Europe as well as America's closest competitor, the Soviet Union. The United States stood alone.

Even the Soviets admitted as much, conceding to the "indisputable superiority" of the United States during the 1950s. America, noted Sergei Khrushchev, son of the Soviet Premier, had "ringed the Soviet Union with airbases. American strategic bombers were capable of turning any Soviet city into another Hiroshima or Dresden." And there was nothing the Russians could do to stop them if America was so inclined.[1]

But most Americans weren't. They felt secure in their dominance and didn't want more war—nor did much of the rest of the world. With the level of carnage witnessed over the previous three decades, people everywhere wanted peace. Even though both wars had not really touched the American homeland, most Americans, especially those in Congress, stood with much of the rest of the globe. They had no stomach for more warfare and little eagerness to spend money on weapons designed for the next conflict. The development of American atomic weapons continued unabated, including the more lethal hydrogen bomb, for use as the most effective of deterrents, as did the military occupation of Germany and Japan, but most of the country concerned itself with internal matters. America prospered in the years after the war, focusing its industrial capacity on manufacturing domestic products such as cars, refrigerators, air conditioners, televisions, vacuum cleaners, dish washers, and washing machines. International trade was just 3 percent of GDP. Swords were being beaten into plowshares.

Not so in the Soviet Union. The Russians had other ideas from the start. Their leadership "feared another catastrophe and did everything possible to avoid it," wrote Sergei Khrushchev. They were not about to let the United States remain in the catbird seat without a fight. After the "Great Patriotic War" and their horrific losses, the Soviets never took their foot off the gas pedal, pursuing an aggressive foreign policy of both propagandistic persuasion and Red Army conquest, continuing military production centered on their own nuclear weapons program but, more importantly, on the development of rockets and missiles to deliver those catastrophic warheads anywhere in the world.[2]

Like the Americans, the Soviets had grabbed a number of German rocket engineers after the collapse of the Third Reich but were not utilizing them to the fullest extent possible. They didn't need to, for they had their own rocket genius in Sergei Korolev, who ran the Russian missile program. Korolev's true identity was completely unknown to American intelligence agencies at the time. The CIA knew he existed but had no clue who he really was, referring to him as the "Chief

Designer." The Soviets kept him a well-guarded secret, fearing that if the Americans found out who Korolev was, they might try to kill him. He was so important to Soviet efforts in rocketry and space exploration that protecting his identity was a national security imperative. Whereas the United States would come to rely heavily on German technology and ingenuity, the Soviets relied on Korolev because "he thinks that he can do it better than Germans. So Germans in the Soviet Union [were] more observers of what the Soviets were doing, but not participants in the project," recalled Sergei Khrushchev, who was himself a Soviet rocket engineer.[3]

Instead of rockets, the United States placed its faith in long-range bombers to deliver its nuclear ordinance, as it did at Hiroshima and Nagasaki, the very events that shook the Soviet leadership. America's missile teams, including its own German engineers, were not moving swiftly toward new systems to launch bombs, or even satellites, because of a general reluctance in Washington. With so many long-range bombers and the overseas bases to support them, why spend great sums of money on missiles? But the Russians were working feverishly with the full backing of the Kremlin.

As far as Moscow was concerned, the Soviets had already been in a technological contest years before the space race kicked off, while Americans, entering the sleepy 1950s, didn't seem to realize or much care that a competition for space was actually ongoing. With the election of General Dwight D. Eisenhower as president in 1952, the former Supreme Allied Commander in Europe who had organized the defeat of Nazi Germany, Americans were content and felt safe with grandfatherly Ike at the helm. But global realities soon intervened to shake the American people from their slumber. Relentless Soviet pressure would end the peace Americans felt in their security and bring forth a new "Red Scare." And that reality was brought home in October 1957, in what John F. Kennedy later called "the most significant event that took place in the fifties." It was called "Sputnik," and it proved American sluggishness and Soviet aggressiveness.[4]

■   ■   ■

Translated as "Fellow traveler," Sputnik was a 184-pound satellite about the size of a large beach ball. On a crisp fall night, as many Americans watched the premier of *Leave It to Beaver*, a massive Soviet rocket pushed Sputnik into space. It reached an apogee, or high-point in altitude, of 560 miles and orbited the globe every 96 minutes, emitting a consistent "beep" sound that could be picked up by any American with a radio transmitter. It would remain in space for 92 days, making 1400 revolutions of the earth. Though lacking in scientific instruments, that beeping sound was the most terrifying noise imaginable, for it proved two very important things. The Russians, of all people, had reached space first. But more importantly, if the Soviets could launch a satellite, then there was nothing to stop them from launching a nuclear bomb on an intercontinental trajectory and annihilating American cities with massive death and devastation with one push of a button. The age of M.A.D., Mutually Assured Destruction, had begun. The shocking news awoke Americans like nothing else could. Senator Henry "Scoop" Jackson announced on the floor of the U.S. Senate a "National Week of Shame and Danger," while one high-ranking U.S. military official said Sputnik was the "Pearl Harbor of the Technology War."[5]

Sputnik certainly gave Soviet Premier Nikita Khrushchev new opportunities to boast and bluster about what he believed would be the inevitable triumph of communism over capitalism. Americans no longer held such a vast technological advantage over the Russians. The Soviets "were the first to launch rockets into space; we exploded the most powerful nuclear devices; we accomplished those feats first, ahead of the United States.... Our accomplishments and our obvious might had a sobering effect on the aggressive forces of the United States," he wrote in his memoir, for the Americans and their allies "had lost their chance to strike at us with impunity."[6]

The nation's top Democrat, Senate Majority Leader Lyndon Johnson of Texas, believed, like other American officials, that Sputnik might lead to war. That had to be the Soviet intention, he thought. "Soon," he said, "they will be dropping bombs on us from space like kids dropping rocks onto cars from freeway overpasses." He was at his ranch in the Texas Hill Country when he received the news about the Soviet satellite. After dinner with guests that night, he took a walk along the Pedernales River. He was in "profound shock" as he pondered a frightening thought, that "it might be possible for another nation to achieve technological superiority over this great country of ours." America, "we all comfortably believed, was the world's most advanced nation in science and technology." Now that seemed to be in jeopardy.[7]

Johnson chaired the Senate Armed Service's Subcommittee on Preparedness and immediately scheduled hearings for November to look into the Soviet feat. But this was not a knee-jerk reaction by LBJ; he had long urged the U.S. military to get more active in rocketry because they were "moving too slowly." The United States had to escape what he called "the bonds of inattention and inaction that had gripped the 1950s." In the years after World War II, Johnson had pushed the Pentagon to "pool our best scientific brains on guided missile research, just as we pooled our best brains to develop the atomic bomb."[8]

One of his top aides believed his aggressive stance on this most serious matter could also help him politically, vaulting him to the top of his profession, a goal Lyndon Johnson had dreamed of since childhood. George Reedy sent him a memo on October 17, nearly two weeks after Sputnik, advising Johnson that the issue, "if properly handled, would blast the Republicans out of the water, unify the Democratic Party, and elect you President. You should plan to plunge heavily into this one. As long as you stick to the facts and do not get partisan, you will not be out on a limb." Though LBJ seemed always to have politics on his mind, this was not a political issue to him, or at least not totally, but, for the most part, one of national security.[9]

Among those across the country who, like Johnson, were shocked and fearful, were members of the news media. When word of the Soviet feat came across the wires at NBC, correspondent Morgan

Beatty, who had covered World War II in the field, excitedly exclaimed, "The Russians have put a satellite in Earth orbit! They've been talking about it, and damn it, they've really done it!"[10]

Word of Sputnik also stunned a cautious Eisenhower administration that had little appetite for spaceflight. Ike's Defense Secretary, Charles E. "Engine Charlie" Wilson, as well as his deputy, Donald Quarles, wanted no part of what they considered little more than fantasy. Eisenhower's attitude was pretty much along the same lines, not to mention his concern about the costs involved, which would presumably be fairly hefty. Ike's chief of staff, Sherman Adams, a former governor of New Hampshire who also possessed a low opinion of spaceflight, jeeringly remarked that the administration was "not intent on attaining a high score in any outer-space basketball game." The president's advisor on foreign economic policy, Clarence Randall, called the satellite a "silly bauble." These comments reflected administration attitudes over the course of the previous year, when they had gone so far as to impede every effort to put a U.S. satellite into orbit ahead of the Russians. America's chief rocket scientist, Dr. Wernher von Braun, had the capability of launching an American satellite at least a full year *before* the Soviets, and was poised to do so with his Jupiter-C missile, originally designed as a weapon to carry a nuclear payload.[11]

Working under the auspices of the Army, von Braun and his team had added a fourth stage to the Jupiter missile that would have provided the appropriate power to boost the nose cone of the Jupiter up to 17,500 miles per hour, the speed needed to overcome gravity and remain in orbit. The launch was scheduled for September 1956, more than a year before Sputnik. Von Braun knew that the administration did not want an American satellite in orbit, at least not yet. But he decided to launch it anyway, on his own authority and under the guise of a simple missile test, and then claim that the satellite had gone into orbit by accident. The commander at Cape Canaveral, Lieutenant Colonel Asa Gibbs, got wind of the plan and ordered the rocket back to the hanger. The military also filled the crucial fourth-stage with sand to keep the nose cone, which did not contain any scientific instruments, from reaching orbit whenever the test was to be conducted. Von Braun's efforts were stopped so as not to antagonize the Russians and further exacerbate Cold War tensions.[12]

Von Braun's rocket launch was eventually a success, even though it did not reach orbit. It climbed to an altitude of 682 miles out into space, flew 3,300 miles downrange, and reached 16,000 miles an hour. The test certainly got the attention of the Russians, who feared America was poised to launch a satellite, so they threw their efforts into hyperdrive in order to get there first. And they did, causing Soviet Premier Nikita Khrushchev to boast that America "sleeps under a Soviet moon." Now with the Communists in space before the Free World, the United States was reeling from the shock, and Ike was feeling the heat for allowing the Soviets to get ahead of America.[13]

On the night Sputnik went into orbit, von Braun was entertaining officials from Washington with dinner and drinks, a group that included Neil H. McElroy, who would soon be the new secretary of defense. McElroy was much more in favor of spaceflight than his predecessor, Charlie Wilson. Von Braun was in the midst of briefing McElroy on what his Army rocket team could do to put America into space first when his public affairs director, Gordon Harris, brought him the news that the Russians had "just announced over the radio that they have successfully put up a satellite." Von Braun was unhappy but wasn't all that surprised. "We knew they were going to do it," he told McElroy. "They kept telling us, and we knew it."[14]

Eisenhower had, at least at first, joined those in his administration who downplayed Sputnik, calling it just "one small ball in the air, something which does not raise my apprehensions, one iota." Charlie Wilson also tried to calm public fears, hitting directly at comments by Lyndon Johnson. "Nobody is going to drop anything down on you from a satellite while you are sleeping, so don't start worrying about it," he said. Both were true statements, but Eisenhower well understood why the American people were scared. And the media weren't helping matters much. "Sputnik is a serious threat—if not to our immediate security, then to our sense of security," said Walter Cronkite of CBS News. Eisenhower eventually admitted that the Soviet achievement was "impressive," he wrote in his presidential memoirs. "The size of the thrust required to propel a satellite of this weight came as a distinct surprise to us. There was no point in trying to minimize the accomplishment or the warning it gave that we must take added efforts to ensure maximum progress in missile and other scientific programs." And he knew just who could do it.[15]

■   ■   ■

Though it seemed to just about everyone with half a brain that the United States was behind at this point, it had an ace up its sleeve in von Braun, the German rocket genius and mastermind of the Nazi V-2 rocket that had struck fear into the hearts of British civilians during World War II. Von Braun had been enamored by rockets and space travel since he was a young boy, voraciously reading authors such as Jules Verne. He joined the early German rocket team under Dr. Hermann Oberth at the Technical University in Berlin in 1930 before the Nazi takeover of Germany. The German military had already placed a premium on rocketry because of the restrictions of the Treaty of Versailles, which greatly limited the German military but did not include rockets and missiles, owing to their primitive and largely unknown nature in those early days. Work continued under the Nazi regime and the only way scientists such as von Braun could continue was by joining the party. And join von Braun did, eventually becoming a member of the SS. For him it was a "Faustian bargain," but he was never an ideological Nazi. In fact, he was arrested by the SS and detained for a time because of suspicions the party held regarding his loyalty. An appeal to the German high command won his release. But despite his task at hand, rockets for warfare, he still dreamed of using the technology for space travel. His enthusiasm was so great that his superiors warned him not to speak of such topics in front of the high command, especially Hitler.

Yet the success of the V-2 had a silver lining. Ironically, "German ballistic missile technology—built to kill people—laid the foundation for spaceflight," wrote Douglas Brinkley.[16]

With the end of the war and the destruction of Hitler's Reich, von Braun came to the United States via Operation Paperclip, a military intelligence project that grabbed what Walter McDougall called "the most prized spoil of the war": the best German rocket engineers, as well as all their plans and rocket parts, and brought them to America. The Soviets had taken some but the United States got by far the best of the lot, as von Braun and his entire team decided days before V-E Day that living in a free America was far preferable to another totalitarian dictatorship in Russia. To pull it off, von Braun hid dozens of crates full of rocket plans and missile parts in a mine in southern Germany to keep them from falling into the wrong hands. Russia and the SS had both sent agents in hot pursuit of the rocket team. Escaping the dragnet, von Braun eventually made it to American lines and lived the rest of his life in freedom.[17]

Von Braun and his team constituted what has been dubbed "the most remarkable and enduring group of engineers the world has ever seen." Employed by the U.S. Army, first at Fort Bliss, Texas, then later at the Army rocket range in Huntsville, Alabama, their task was to develop missiles to deliver nuclear warheads. But even though he was forced to build weapons of war for the Nazis, and now the Americans, von Braun had no real interest in such things. For von Braun, working on rockets with military application was a means towards realizing his dream of launching mankind into space. He had been dreaming of spaceflight for many years, even trips to the moon and Mars, publishing numerous academic papers throughout the 1950s laying out detailed plans for future travel. It was these great dreams of the future that produced his friendship with Disney.

But the lack of American resolve prevented von Braun from getting closer to achieving his far-reaching goals. While the Soviets blazed ahead in rocketry, the Americans sat idly by. "The United States had no ballistic missile program worth mentioning between 1945 and 1951," von Braun admitted. "These 6 years during which the Russians obviously laid the groundwork for their large rocket program, are irretrievably lost."[18]

American reluctance to pursue rocketry was nothing new but, in fact, went back decades. An American physicist, Robert H. Goddard, was an early visionary and pioneer of the possible use of rockets. Even though von Braun and his team of German rocket scientists get all the glory these days, Goddard was at the forefront of the study of rocketry and space travel, and not only as a theorist but as a man of action. Fascinated with the cosmos since his early days as a sickly child, Goddard was bitten by the space bug after reading H. G. Wells's famous novel, *The War of the Worlds*, published in 1897. He eventually earned a Ph.D. in physics, became a professor at Clark College in Massachusetts, and began to work out theories on rocketry and space travel. He believed spaceflight was possible, that rockets could function in the vacuum of space, that liquid fuels, such as hydrogen and oxygen, would be superior to solid fuels like gunpowder, that cylindrical combustion chambers were the way to go in rocket construction. Ridiculed by the press for what seemed like absurd theories of sending rockets to the moon, Goddard nonetheless pressed on and built his first liquid-fueled rocket in 1926. Although it rose to only 41 feet in altitude, Goddard established himself as an early founder of the science of rocket flight, eventually owning 241 patents, including one for a multi-stage rocket. Before his death in 1945, he had tested dozens of rockets in the open spaces near Roswell, New Mexico, with the most successful traveling to a height of 9,000 feet in 1937.[19]

Goddard's theories would be studied and expounded upon for years to come, even by early rocket scientists in Russia, who praised his work in what has to be the irony of ironies. "Goddard became respected, even idolized, in the Soviet Union," writes Douglas Brinkley. Von Braun was also quick to credit the American for his later success. "His rockets...may have been rather crude by present-day standards, but they blazed the trail and incorporated many features used in our most modern rockets and space vehicles." His "experiments in liquid fuel saved us years of work, and enabled us to perfect the V-2 years before it would have been possible," he said. "Dr. Goddard was ahead of us all." The Germans took Goddard's work, which Goddard himself thought they had stolen from him, and forged ahead while America was uninterested, not then and not after World War II.[20]

■   ■   ■

After Sputnik, America's attitude began to change quickly, though the new resolve may not have been readily apparent to the general public. On October 8, 1957, four days after Sputnik, Neil H. McElroy, a proponent of an American space program, was officially named the new secretary of defense. Although it was announced as a simple retirement of longtime official "Engine Charlie" Wilson, there can be little doubt that the move had everything to do with the need for a more aggressive posture in space policy. On the same day, Ike held a conference with some of his top advisors to discuss Sputnik and the possibility of putting an American satellite in orbit soon. A classified memo, made public in 1976, detailed the conference. With the shuffle at the top of the Pentagon hierarchy, Deputy Secretary Quarles, a communications engineer by trade, led the meeting. The president asked Quarles specifically about a report that indicated that America could have put a satellite in space ahead of the Russians with use of the Army's Redstone, an obvious reference to von Braun's Jupiter test the previous fall. Quarles acknowledged the report, yet told Ike that the president's own Science Advisory Committee felt that any U.S. satellite should "proceed separately from military development" in order to "stress the peaceful character of the effort." With no scientific instruments on board, a nose cone in orbit by the Army rocket team would have looked militaristic. While it's likely Eisenhower knew all about the early American effort and probably had as much to do with stopping von Braun as anyone else, he chose to remain coy. Despite the lengthy discussion that day, no hard decisions were made.[21]

While America was considering what to do next, the Soviets struck again, one month later with Sputnik II, a satellite that weighed over a thousand pounds. Only this time it was more than just a satellite; it was essentially a spacecraft, for the Russians boosted a dog, a live animal, into orbit. Named Laika by the Soviets, and "Muttnik" by the American press, the 13-pound mixed-breed canine orbited the earth for several days before dying from lack of oxygen. The satellite itself later burned up in earth's atmosphere in April 1958 after remaining in space for 150 days, making 2,370 orbits. As tragic and unsettling as the Laika story was, the feat itself was still just as shocking to America and the world. Heavier payloads meant more powerful rockets, which could carry larger warheads, and perhaps even a spacecraft for orbiting a man. There were serious questions on the minds of many Americans: Could the Russians be beaten? Some were already concluding that the Soviets possessed an insurmountable lead and the whole American effort should be abandoned. Why waste huge sums of money on a lost cause?

To Americans on the other side of the political divide, Ike seemed to fiddle while Rome burned. Democrats wasted little time hammering the Republican president about his reluctance. He loved to play golf, and his political opponents used that against him. The Democratic governor of Michigan, G. Mennen Williams, wrote a poem attacking the president over Sputnik:

> Oh little Sputnik, flying high
> With made-in-Moscow beep,
> You tell the world it's a Commie sky
> and Uncle Sam's asleep.
> You say on fairway and on rough
> The Kremlin knows it all,
> We hope our golfer knows enough
> To get us on the ball.[22]

Democrats might have sought to make political hay out of the crisis, as any political party would be wont to do, but Eisenhower, then in his second and final presidential term, was not concerned with politics; he had a very full plate of "steadily mounting crises and pressures" in the fall of 1957, something most people never stop to consider. After dealing with the potentially explosive Suez Crisis that ended that spring, the president faced an "alleged interservice rivalry, guided missiles, Sputnik II, Little Rock, Syria, and a variety of other problems," he wrote to his former naval aide. The "Little Rock" reference referred to the ongoing crisis to integrate public schools in the South, based on the 1954 *Brown v. Board of Education* decision by the U.S. Supreme Court. Near the end of September, Eisenhower sent in troops to enforce the court order. To his older brother, Arthur, he wrote, "When I wake up in the morning I sometimes wonder just what new problem can possibly be laid on my desk during the day to come; there always seems to be an even more complex one than I could have imagined."[23]

Although the president had been reluctant to get into the space race, Eisenhower never considered panicking or quitting an option. He understood the global implication of what the Russians had done and knew he needed to match the Soviets fast, despite prodding from Democrats. But he also understood that despite Khrushchev's boasting to the contrary, Sputnik itself did not threaten the American people, at least not militarily. American security was assured. But much of the American public, including Democrats and members of the media, would never see it that way. On October 9, at a press conference, the president fielded many questions, all of them related to Sputnik. And the first question right out of the gate set the hostile tone of the session. "Mr. President, Russia has launched an earth satellite," Merriman Smith of the United Press International began. "They also claim to have had a successful firing of an intercontinental ballistic missile, none of which this country has done. I ask you, sir, what are we going to do about it?"[24]

That was the big question, and Eisenhower's lengthy response did little to answer it. And while the Eisenhower administration failed to mollify the public's concerns, American labors to quickly counter the Soviets were met with abject failure. Rocket after rocket blew up prematurely—if they launched at all—and in front of the press, dignitaries, and military brass, no less. At least part of the trouble was American competition. Even though this was a government-run operation, which would seem to guarantee a monopoly, three U.S. military branches fought over the rights to the nation's missile and rocket program and the funding to go along with launching the nation's first satellite: the Navy, Air Force, and Army. Each branch boasted programs in varying stages of development. The Air Force had the Thor and Atlas rockets, the Army had its Redstone and Jupiter, while the Navy had its Vanguard program. Ultimately the U.S. government chose to award the Navy the task of sending a satellite into space, thinking that under the Navy's stewardship the mission would seem less menacing and militaristic than the under the Army's auspices.

Yet the administration knew that Vanguard was behind schedule and was responsible for the expensive fireworks shows that could be seen on a routine basis down at Cape Canaveral. Von Braun knew as much, telling McElroy on the night Sputnik launched, "You know we're counting on Vanguard. The president counts on Vanguard. I'm telling you right now Vanguard is months away from making it."[25]

And von Braun was right. Vanguard was a mess. One memorable Vanguard flight took place on December 6, 1957, a little more than a month after Sputnik II. It would carry a simple, three-pound satellite, only about the size of a grapefruit, which compared to Sputnik was quite the joke. But the launch, or lack thereof, lifted the rocket just four feet off the pad before it exploded in a massive failure. The satellite managed to escape the inferno and rolled hundreds of feet away and into the bushes, and began emitting its signature beep, prompting the columnist Dorothy Killgallen to ask later, "Why doesn't someone go out there, find it, and kill it?" NBC space correspondent Jay Barbree wrote that it was a "black day for a proud country." The New York Times headline read: "Washington Humiliated." Other newspapers blared mocking headlines around the country and the world: "Rearguard," "Kaputnik," "Stayputnik," and "Flopnik" were among the most colorful. Lyndon Johnson was equally appalled. "The Soviet success and the Vanguard failure

depreciated our prestige," he wrote, noting that "this second-best position was the possible result of our shortsightedness and complacency."
Foreign nations saw the same thing. To a visiting American official,
future British foreign secretary George Brown asked, "Do you Americans
really know what you are doing?"[26]

It surely didn't look like it. American efforts seemed as inept as Soviet
efforts were competent. The Air Force did successfully launch America's
first ICBM, the Atlas, the same month as Sputnik II, an important military objective, but that hardly mattered, not with the space race anyway.
The first flight of the Army's Mercury Redstone in 1960 was an even
bigger embarrassment, lasting all of two seconds, lifting the rocket just
four inches off the ground. Known as the "four-inch flight," the escape
tower jettisoned and landed a few hundred yards away on the beach,
while the parachutes deployed and nearly pulled the fully-fueled missile
to the ground. It easily could have exploded in a massive fireball. Nikita
Khrushchev even got in on the fun and games, tauntingly asking if the
United States needed the assistance of Soviet rocket scientists to get their
space program off the ground.[27]

But the president wasn't unnerved and remained steady during the
crisis. According to the Sputnik memo in early October 1957, Quarles
told Eisenhower that the Army could put up a satellite within four months,
a full month faster than what the Navy was predicting. Eventually Ike
gave von Braun and the Army missile team the go-ahead to put an American satellite into orbit. And not only was von Braun as good as his word;
he was faster than anyone in the administration realized. On January 31,
1958, less than four months after Eisenhower's green-light, Explorer 1
lifted off from Cape Canaveral via a Jupiter rocket. The satellite, weighing
little more than thirty pounds, was boosted into orbit. It also contained
several scientific instruments, as a way to show that this was a peaceful
effort, including one to study and measure the radiation in earth's orbit.
The experiment was under the direction of Dr. James Van Allen of the
University of Iowa, who had previously theorized about the existence of
radiation belts ringing the earth. After confirmation by a later satellite,

the Van Allen radiation belts now bear his name. Explorer 1 would transmit data back to the ground for several months and remained in space until 1970. Two months later, in March, Explorer III and Vanguard III joined Explorer 1 in orbit.[28]

"It was one of the great moments of my life," von Braun would later say of that first launch. "I only regret we weren't permitted to do it earlier." Though many Americans, especially those around Huntsville, were a little suspicious of the former Nazi SS man upon his initial arrival, von Braun eventually became an American citizen, a celebrated hero, and, as far as Huntsville was concerned, one of their own. In recognition of the new American's effort, Eisenhower awarded von Braun the Distinguished Federal Civilian Service Award at a white-tie dinner at the White House.[29]

Under von Braun's direction, America quickly surpassed the Soviet Union in launching satellites into space, at least as far as the number of satellites was concerned. But the Soviets continued to one-up America every chance they got. In May 1958, the Russians launched Sputnik III, a spacecraft that weighed an astonishing 2,900 pounds and carried a dozen scientific experiments, as many as the next six U.S. satellites combined. Premier Khrushchev derisively referred to American birds as mere "oranges."[30]

■   ■   ■

Launching satellites was certainly not the goal of either the Russians or the Americans, though it marked an important benchmark. With Laika the dog in its hold, Sputnik II had proved that the Soviets were at the very least curious about manned spaceflight, presumably with the ultimate objective of a lunar landing. In 1959, the Russians launched three lunar probes—two lunar flybys, one of which snapped photos of the never-before-seen far side of the moon, and one mission that landed a probe on the lunar surface.

In January 1958, Senator Lyndon Johnson had completed his hearings on what the *New York Times* was now calling the "Space Race." In the committee's report, Johnson made bold declarations of the power nations could wield from space. "Control of space means control of the world. From space, the masters of infinity would have the power to control the earth's weather, to cause drought and flood, to change the tides and raise the levels of the sea, to divert the gulf stream and change temperate climates to frigid," he wrote, eager to play on people's fears. "In essence, the Soviet Union has appraised control of space as a goal of such consequence that achievement of such control has been made a first aim of national policy. [In contrast], our decisions, more often than not, have been made within the framework of the Government's annual budget. Against this view, we now have on record the appraisal of leaders in the field of science, respected men of unquestioned competence, whose valuation of what control of outer space means renders irrelevant the bookkeeping concerns of fiscal officers."[31]

If America was to have any hope of catching the Soviets and eventually overtaking them in space, things had to change. Washington decided to restructure its efforts with a new organization. Until 1958, most everything having to do with flight, at least as far as the federal government was concerned, had been conducted under the National Advisory Committee on Aeronautics (NACA), which was founded on March 3, 1915, to conduct aeronautical research, as primitive as it was in the early days. After World War II, it moved into rocket planes, like the X-1, in an attempt to break the sound barrier, shatter speed records, and fly to the edge of space, utilizing its own test pilots, one of whom was a twenty-eight-year-old civilian named Neil Armstrong. By the late 1950s, NACA was testing the X-15, capable of speeds in excess of Mach 6 and altitudes of up to 67 miles, the operational limits of a manned airplane. NACA had 8,000 total employees and some of the best engineers around, with 5 laboratories, including the Jet Propulsion Laboratory at Cal Tech, the Vanguard program, and the Army rocket team headed by von Braun.[32]

Once flying in space was set as the new objective, everything would be folded into a new independent agency, the National Aeronautics and Space Administration, or NASA, tasked with space exploration "for the benefit of all mankind." This new administrative body was a fruit of Lyndon Johnson's labors, as well as other members of Congress. NASA was to be a civilian agency, not a military one. It wouldn't be headed by a general but a civilian administrator appointed by the president to oversee "non-military space efforts." Eisenhower signed the bill in July 1958, and the agency opened for business in October. As an independent civilian agency, NASA would not fall under any cabinet secretary but would answer directly to the White House. And NASA would get everything it needed to do the job at hand. Even von Braun's Army rocket team was eventually transferred to NASA at the end of 1959. The transfer was only logical. At that point, von Braun's team was working on producing large boosters to be used for space purposes, not warfare.[33]

President Eisenhower chose T. Keith Glennan to serve as NASA's first administrator. Glennan, then president of the Case Institute of Technology, had also worked with the Atomic Energy Commission and the National Science Board. In his first month, Glennan chartered the Space Task Group (STG) at Langley, Virginia, which would be headed by Robert Gilruth, a NACA engineer and a man well respected in aviation. Christopher Kraft, also an engineer and pioneer of NASA, has called Gilruth "a father of human space flight." The STG's main task was to manage NASA's first project, Mercury, conceived in October 1958, with an objective to put a man in space.[34]

NASA's headquarters would be in Washington, D.C., led by the administrator, a deputy administrator, and an associate administrator. The new administration would manage the entirety of the space program, determine space policy, obtain funding from Congress, and promote the goals of the agency before Congress and the country as a whole. It would encompass and oversee a number of groups and sub-agencies. The Redstone Arsenal at Huntsville, where von Braun and his team were cooking up their rocketry experiments, became the George C. Marshall

Space Flight Center, with 4,600 employees. The Marshall Center would be responsible for designing, testing, and launching rockets, as well as supervising their production at various plants around the country. Von Braun would be its director. Their first task was to produce a new rocket system named the Saturn. Meanwhile, other bases around the country were repurposed to meet the requirements of the space program. For instance, the Goddard Space Flight Center in Maryland, along with JPL in California, would be responsible for managing vast amounts of data, tracking the spacecraft, and communications from the spacecraft to the ground.[35]

Despite the massive effort the Americans had set in motion, in August 1960 the Soviets managed to shock the world again, launching a capsule with two dogs, Belka and Strelka, aboard their new Vostok spacecraft. This time the animals survived twenty-four hours in orbit and returned to earth safely on August 20. Both dogs were retired after the flight. One later gave birth to a litter of puppies, one of which was given to Jackie Kennedy as a gift.

With the Vostok launch, the Russians made their target of putting a man in space clear. If recent history was any indication, it was a good bet that they would soon prevail in that endeavor. NASA and Project Mercury were both still in infancy, and the Russians looked much closer to sending a man to space than their American rivals. By this point, everyone knew that the space race was a new front in the Cold War between the United States and the Soviet Union. And in the eyes of both competitors, the space race was not just another competition for two rival powers to engage in on the world stage; it was perhaps the most important competition to win. The legitimacy of each regime was at stake, as the Cold War was as much about which system could produce superior technology as it concerned political and economic principles. "Both the Soviet Union and the United States believed that technological leadership was the key to demonstrating ideological superiority," Neil Armstrong recalled. "Each invested enormous resources in evermore spectacular space achievements. Each would enjoy memorable successes.

Each would suffer tragic failures. It was a competition unmatched outside the state of war."[36]

But to prevail, the United States would need a fresh perspective that would push American technology forward. And that would take a new president, young, energetic, and with a vision, along with eager military test pilots courageous enough to climb on top of what was essentially a bomb to be launched into space. Together, the president and those pilots would put the United States on course to make it to the moon.

I

THE BEGINNING: MERCURY

Captain Gus Grissom of the United States Air Force sat uncomfortably as he faced a swarm of reporters and photographers packed into NASA's temporary headquarters in Washington, D.C. The date was April 9, 1959. As a career military man, Grissom was unaccustomed to the public eye. In fact, it was far from a welcome sight, even for a skilled pilot and engineer who had flown one hundred combat missions during the Korean War. Facing enemy MIGs was one thing; facing more than two hundred members of the media snapping endless pictures and seeking to probe your innermost thoughts was quite another. But as nerve-racking as the experience might have been, Gus wouldn't have traded it for any other assignment; it was where he wanted to be—at the top, sitting among the final seven of an elite group, America's first astronauts. And they were in the nation's capital to be introduced to the American people and the world.

The public was eager to learn more about the first men whom the space agency had chosen to ride a rocket into the void. Space fever had begun to take hold across the United States, and the seven astronauts would quickly become national heroes. Seated alongside Gus were Navy

Lieutenant Malcolm Scott Carpenter, Air Force Captain L. Gordon Cooper, Marine Major John H. Glenn, Navy Lieutenant Commander Walter M. Schirra, Navy Lieutenant Commander Alan B. Shepard, and Air Force Major Donald K. Slayton. Several of the newly minted astronauts would later admit that their newfound celebrity terrified them. But public exposure was a price they were willing to pay to fly in space, an endeavor for which they showed no fear whatsoever.

As flashbulbs popped and correspondents jockeyed for position, Gus and his six colleagues, all distinguished military test pilots, sat behind a long table covered with a felt cloth and lined with microphones. Steely-eyed and serious, the Mercury Seven, now seen as national heroes, would take on the enormous and very dangerous task of countering the Soviets in what seemed even at this early stage like their almost insurmountable lead in the space race. But that didn't matter to the media, who stood in awe of these seven brave souls. The Mercury Seven, wrote journalist James Reston, "talked of the heavens the way old explorers talked of the unknown sea." And they shared the same sense of adventure to explore that unknown. Their objective was to win the race for space and possibly the moon, and the soft-spoken man known as Gus would prove to be an integral part of the equation.[1]

■   ■   ■

Gus was born Virgil Ivan Grissom on April 3, 1926, in Mitchell, Indiana, the oldest of Dennis and Cecile King Grissom's four children. He had two brothers, Norman and Lowell, and one sister, Wilma. Years later a colleague misread Gris as "Gus," and the nickname stuck. In his adolescent years, Gus's father, Dennis Grissom, held a job with the Baltimore and Central Railroad, which afforded the Grissom household some comfort in the midst of the Great Depression. "I worked six days a week at 50 cents an hour," Dennis Grissom remembered. "Men got laid off all around me. I worried I'd be next, but at $24 a week, my family was well off." At least comparatively speaking. "We were far from rich,"

Gus remembered, but "we had a warm, comfortable family life, strongly reinforced by our parents' deep religious convictions."[2]

Gus's father never pushed his sons to follow him in the railroad business and encouraged them to follow their own dreams. "I suppose I built my share of model airplanes," Gus recalled about his childhood, "but I can't remember that I was a flying fanatic." He also readily admitted that he was not "much of a whiz in school," a "case of drifting and not knowing what I wanted to make of myself. I'm reasonably certain that most of my teachers in high school didn't think I'd make it to college, or if I did, be able to keep up with the grades." But he was active in his teenage years, played high school baseball, was in the Boy Scouts, and ran a morning and evening paper route. In the summertime, he picked fruit in the many orchards around town to put a few extra bucks in his pocket.[3]

While a sophomore in high school he met an incoming freshman named Betty Moore, whom he would later marry. Gus grew up in town, a "city boy," as Betty described him, with a comfortable enough lifestyle to have indoor plumbing, a rare commodity at the time, especially for those who lived in the country, where Betty grew up. But those differences mattered not to Gus. "The first time I saw you I decided you were the girl I was going to marry," he would later tell her.[4]

As sparks started to fly between Gus and Betty, World War II came to America with the Japanese attack on Pearl Harbor. Gus was 15, the age when war and the daring feats of soldiers, sailors, and aviators most strongly pique the curiosity of young men. He had already been reading about the great air duels in Europe between the Luftwaffe and the Royal Air Force Fighter Command over the past year and made up his mind to join the war effort once he graduated from high school in 1944. He decided on the Air Force, then a part of the Army, for a simple reason: "flying sounded a lot more exciting than walking," he would later write. Gus's orders came in August 1944, and he was soon off to Texas for training, hoping he could qualify for aviation. But the war began to wind down before Corporal Gus Grissom could receive any flight training.[5]

Mustered out of the service after the war ended, Gus married Betty back in Mitchell on July 6, 1945, and the new couple began their life together. The early going was tough. With millions of young men returning home from the front, jobs were hard to find and good jobs rarer still. Living in a small apartment on Main Street in Mitchell, Gus worked for Carpenter's Bus Body Works putting doors on school buses, while Betty took a job at Reliance Manufacturing, a factory that made shirts for the Navy, a position she had held the previous summer. Gus took numerous days off from work and drove around the area looking for a better job but had no luck. He was edgy and impatient, wanting more out of life than what he had in Mitchell. "He got on my nerves a little bit in those days," remembered Betty. "I really didn't mind working. In fact, I enjoyed it. It gave me something to do while he was running around looking for something else. We weren't making much money, but we had grown up that way, and when you're young, you can adapt to a lot of things. Whenever he was grumpy, I tried to laugh it off. I said: 'I can't get upset with you because my mother said if we had a fight I couldn't come home.'"[6]

But Gus knew what he wanted out of life—flying jets—and he also understood what he had to do if he wanted to fly for the U.S. military. "I realized soon after I got into the Air Force…that I needed more technical training if I was going to get ahead," he wrote in *We Seven*. He needed a college education so he could return to the service as an officer and begin training as a pilot. So, he began looking through engineering course books from Purdue University, which gave Betty "a distinct sense of relief," although she didn't say anything to him about it. Then one morning over breakfast he asked her, "What would you say if I decided to go back to school?" Being the supportive wife that she was, Betty would stick with him. "If that's what you want, it's fine with me," she said.[7]

It wouldn't be quite so simple and easy. Deciding to go was one thing; getting accepted and enrolled was something else altogether. He would only qualify for $105 a month from the GI Bill and, with so many vets returning from the war, open spaces in college and universities were minimal. To get the ball rolling he needed a letter of recommendation from his high school, but his "average high school record caught up with him." After a sit-down with the principal, George Bishop, who remembered Gus "as an average solid citizen who studied just about enough to get a diploma," he was able to get a good letter to send to Purdue. He was eventually accepted.[8]

So, in the fall of 1946, Gus moved to West Lafayette, Indiana, to begin a study of mechanical engineering at Purdue University. With hard work and dedication, Gus and Betty forged ahead as a team, doing whatever was necessary. To make it work, they sold their car, and Betty moved back in with her parents when Gus left for school because he had to board in a small, basement room with another male student, not exactly the appropriate atmosphere for a young wife.[9]

But during his second semester, he found a place for the both of them, a simple room in a boarding house where they lived for a year, before finding a "pint-sized apartment near campus." Gus studied and, to make ends meet, worked as a short-order cook flipping burgers thirty hours a week after classes, and Betty got a job as a long-distance telephone operator, working the 5:00 p.m. to 11:00 p.m. shift, which boosted their monthly income to $350. Gus worked through the summers, skipping holiday time, and finished his degree in three and a half years, graduating in 1950, an achievement he readily admitted was made possible by Betty's love, hard work, and never-ending support.[10]

He always knew that his career would not have been possible without a supportive wife at home. Betty was as much a part of his success as he was. She liked to joke that she had earned a P.H.T.—"Putting Hubby Through." Gus was quick to give her credit, even when he reached the top of his profession. At the Mercury introductory press conference in 1959, the new astronauts were asked about the support they had from their family at home. Gus answered that Betty was very supportive "or, of course, I couldn't be here," which drew a chuckle from his fellow fliers and the assembled press. "She's with me all the way."[11]

After finishing at Purdue, and with the "Air Force bee in my bonnet," Gus returned to military service, training first at Randolph Air Force Base in Texas, then Williams Air Force Base in Arizona, while Betty stayed in Indiana. She was later able to join him in Arizona with the couple's first child, six-month-old Scott, in tow. In March 1951, Gus received both his flying wings and his bars as a newly commissioned second lieutenant.[12]

Not long after Gus received his wings, a new conflict would mobilize American forces. War had broken out the previous year in Korea. The Air Force shipped Gus out in December 1951. He flew an F-86 Sabre with the 334th Fighter-Interceptor Squadron. He had orders to remain in Korea "until I had completed one hundred missions—or got knocked out—whichever came first." During the war, Air Force pilots were transported from their barracks to the flight line by buses, and the rule was that no pilot could take a seat on the bus, but had to remain standing during the ride to and from the flight line until he had been shot at by an enemy MIG. Gus took a seat after just two missions. Flying the wing position, Gus's job was to protect the flanks of the other flyers during combat missions and keep a sharp eye out for MIG fighters. After his 100th mission, he had been shot at plenty of times but still had not lost a single plane that he had been tasked with protecting. And more importantly, he hadn't lost his own plane. He was awarded the Distinguished Flying Cross and volunteered to fly twenty-five additional missions but was sent home instead, his assignment completed.[13]

Returning home alive and well, he "served a hitch teaching some younger pilots how to fly," studied aeronautical engineering at the Air Force Institute of Technology at Wright-Patterson Air Force Base in Dayton, Ohio, and completed the Air Force's Test Pilot School at Edwards Air Force Base in California. He then ended up back in Dayton test-flying the newest fighter planes. "This was what I wanted all along," he wrote, "and when I finished my studies and began the job of testing jet aircraft, well, there wasn't a happier pilot in the Air Force."[14]

■   ■   ■

By the late 1950s, there was a new game in town for pilots: space-flight. While Gus was busy test-flying jets, America was at war with the Soviet Union for the high ground of space. The initial battle over satellites was winding down; now the goal was manned spaceflight under the federal government's new agency, NASA. With the creation of NASA in 1958 came Project Mercury, the fledgling agency's first manned flight program. The goal was simple: put a man in space, orbit the earth, come home safely. Nothing more. But since such feats had never been done, or even attempted, nobody thought it was going to be easy. First, though, NASA had to find people brave enough—or dumb enough, depending on one's perspective—to climb on top of a rocket filled with explosive fuel, ride the missile into space, orbit the earth at over 17,000 miles per hour, then plunge back into earth's atmosphere where friction from the increasingly thickening air on the spacecraft would send temperatures into the thousands of degrees, before splashing down in the ocean to be rescued by the U.S. Navy. Not exactly a job that would suit most people.

NASA decided early on that it wanted military test pilots to serve as America's original astronauts, even though quite a colorful list of alterna-tives was put forth in early meetings, including surfers, acrobats, moun-tain climbers, race car drivers, and daredevils. But this was a serious operation, relying on unproven mathematical trajectories and re-entry theories, where men would be flying a number of untested vehicles that sat atop millions of pounds of highly combustible fuel that could explode at any second, as the Cape Canaveral fireworks shows had demonstrated to everyone who cared to watch. America needed the best, and the best could be found in the U.S. military.

To recruit prospective candidates, NASA went to the Pentagon and met with commanders from the Navy, the Air Force, and the Marine Corps, each of whom agreed to participate in the new program even though they might lose at least a few of their very best pilots. This was the Cold War, an all-out battle with the Soviet Union, one of those rare times when the leadership of individual branches of the military were willing to put aside their mutual rivalries in the pursuit of common goals. If the best military pilots were needed to beat the Russians in space, then so be it.

A few career aviators were unsure about leaving the military to join a civilian agency, wondering if they would be seen as deserters, or if they could return to the ranks if things did not work out as astronauts. Many chiefs assured them that they had the full backing of their respective branches. The tough, no-nonsense Air Force chief of staff, General Curtis Lemay, the architect of the firebombing campaign over Japan during World War II, assured his pilots that he supported them 100 percent if they were chosen for astronaut duty. "There are a lot of people who'll say you're deserting the Air Force if you're accepted into NASA. Well, gentlemen, I'm Chief of the Air Force and I want you to know I want you in this program," he told a group of prospective astronauts in 1962. "I want you to succeed in it, and that's your Air Force mission. I can't think of anything more important, so don't any of you feel like a deserter." Lemay's attitude was typical of each military branch.[15]

Of all the pilots in the U.S. military, just 110 fit NASA's bill—58 air force pilots, 47 naval aviators, and 5 marines. Potential astronauts had to meet several basic qualifications: they had to be younger than 40, in good physical condition, under 5'11" tall and not over 180 pounds, possess a bachelor's degree, preferably in engineering, be a graduate of test pilot school, have at least 1,500 hours flying time, and be qualified in jet aircraft. Gus met the qualifications, even though he didn't know it yet.[16]

One day in January 1959, Gus received what could only be described as a strange order. He was directed by the Pentagon to report to Washington. The teletype message was marked "Top Secret." It gave no specifics, just that he had to wear civilian clothes and report to a specific address in the capital. When he came home and told Betty, she made light of the situation. "What are they going to do? Shoot you up in the nose cone of an Atlas?" she asked. Gus had no clue but he would find out soon enough.

First in a room with other military test pilots, each was interviewed separately, then given a classified briefing about Project Mercury, the effort to
put a man in space. Because he "somehow met their requirements," Gus
was "invited" to try out for the new program. Anyone who did not want
to participate in the program could return to military duty with no hard
feelings or negative marks in their service record.[17]

Gus was certain he wanted to give it a shot but he had a few questions buzzing around in his head. Years later, when a friend asked him
why he had decided to join up with Project Mercury, Gus told him that
if "I had been alive 150 years ago I might have wanted to go out and
help open up the West." The urge to explore new frontiers was strong
in him. Yet at least initially, Gus needed to consider a few serious issues.
He had his Air Force career to think about, not to mention a family to
support, especially now that he and Betty had two young boys, Scott
and his younger brother Mark, who was born in 1953. He was doing a
job that he "thoroughly enjoyed," test-flying for the Air Force. If he
should join the astronaut corps for a few years and wanted to go back
to the Air Force at some point, would he lose his seniority and have to
"go back as a green hand?" Was Mercury "a serious research project"
or not? Because it sounded to Gus "a little too much like a stunt instead,"
an opinion many others shared. The reason he and his skeptical colleagues thought it might not be a serious enterprise and something they
might want to stay away from was that there might not be much flying
in Mercury. These new "astronauts" might be nothing more than passengers in a capsule. In fact, a number of military test pilots would turn
NASA down. Gus was almost one of them. He "liked flying too much"
to give it up for this new "stunt" in space. "I had never been much of a
science-fiction or Buck Rogers fan," he admitted. "I was more interested
in what was going on right now than in the centuries to come." But in
the end, the prospect of adventure was just too exciting to pass up. He
wanted to go for it.[18]

Gus recalled the episode later. "I knew instantly that this was for
me. This was where the future of test piloting lay. But they were right: It

was only fair to see how Betty felt, although I think I could have told them then and there." After discussing it, she asked him, "Is it something you really want to do?" "Yes, it is," he replied. "Then do you even need to ask me?" With Betty's agreement, Gus would throw his hat into the ring to join the astronaut corps.[19]

After a preliminary round of eliminations, 32 of the 110 pilots who met the qualifications were selected to travel, clandestinely disguised as civilians, to the Lovelace Clinic in Albuquerque, New Mexico, for extensive physical and psychological testing to determine the best group to conduct space flights. The days of medical tests ranged from the routine to the truly absurd, having nothing whatsoever to do with spaceflight. Endless amounts of time on stationary bikes and treadmills, breathing tests, x-rays, isolation chambers, being submerged in hot and cold water, a battery of mental tests and evaluations, and the humiliation of submitting samples of every known human fluid and secretion. Gus didn't perform very well on the treadmill, he admitted, because his heart-rate shot up to 200 beats per minute, leading the doctors to stop the test. But he endured the heat chamber for hours at a temperature of 135 degrees. Deke Slayton considered all the tests "excessive," while Wally Schirra later wrote that the "physical exams at Lovelace were an embarrassment, a degrading experience," a "case of sick doctors working on well patients" with an "array of tortures."[20]

The media knew as much. At the introductory press conference, one question posed to the Mercury Seven was which test was the worst for them to endure. John Glenn put it best, in his open, light-hearted, folksy manner, which the media and the American people came to love and admire. "That's a real tough one. It's rather difficult to pick one because if you figure out how many openings there are on a human body and how far you can go in any one of them…," pausing as many in the audience began to chuckle. "Now you answer which one would be the toughest for you…" The room roared with laughter. No further explanation was necessary.[21]

Of the thirty-two pilots who were invited to Lovelace, only one did not make the cut: Navy pilot Jim Lovell, who had a bilirubin count that was too high. Usually a temporary condition, Lovell's levels improved, and he would later be selected in the "Second Group" and make four spaceflights during his NASA career, including the ill-fated Apollo 13. The thirty-one remaining candidates were shipped off to Wright-Patterson Air Force Base in Dayton, Ohio, for more examinations, namely psychological exams and stress tests. For Gus, he "did not have the slightest idea what they were trying to prove, but I tried to be honest with them and give the doctors straightforward answers without getting carried away and elaborating too much." To his pal Deke Slayton, the exams seemed completely unnecessary. Experienced test pilots all, they had been test-flying jets for years, some with actual combat experience in Korea. "The fact that I had *survived* should have told them all they needed to know about stress," Slayton later wrote. "What were they supposed to learn from hooking me up to an idiot machine with flashing lights?" The eighteen who made it through Wright-Patterson were eventually pared down to what would be the final seven.[22]

Instant celebrity status came with selection to the elite Mercury Seven group. "How the hell had seven guys suddenly become heroes without doing a damn thing?" Slayton asked. "None of us know a damn thing about being an astronaut. Spaceflight is science fiction. We don't yet have the big picture. We'll learn together. We'll fly together." But the press was all over them all the time, even their wives. Gus called from Washington to let Betty know to be ready for it. "The thing is, we don't know what's going to happen," Gus told her. "It's a good bet you'll be pounced on by the press." This was not welcome news for Betty on a good day; but on this day she had the flu and a fever of 102 degrees. But she did her duty, cleaned up the house, went to the doctor for a shot of penicillin, and ran to the grocery store, where she encountered the first reporter and photographer, who had followed her there. After inviting them back to the house, soon a swarm of press people were in her living room, where she patiently answered their never-ending questions while enduring the flash of a thousand bulbs in a span of several hours.[23]

One of the interviewers was NBC's Jay Barbree. Describing Betty as "most gracious," Barbree asked her, "How do you feel about your husband going into space?"

"The two boys, Scott and Mark, and I have been living with a test pilot," she replied. "I don't really feel flying into space is going to be all that different. We feel it will be risky, but if that's what Gus wants to do, then we're all for it."[24]

But despite the occasional headaches, there were perks to becoming an astronaut, incredibly lucrative perks at that. Because of the heightened fame, every news organization wanted a piece of the action. *Life* magazine, though, the top periodical in the nation, was prepared to pay handsomely for exclusivity, doling out a half a million dollars spread out over four years in order to gain a monopoly on coverage of the astronauts and their families. It worked out to an additional $24,000 annually per astronaut, in addition to their regular pay, which ranged anywhere from $8,300 to $12,700 per year, hardly a bad salary in those days.[25]

Within three weeks of their introduction to America as the nation's first astronauts, Gus and his six comrades began training at Langley Air Force Base in Virginia for the first American spaceflight, engaging in a serious, yet friendly, inside competition to determine which of them would be the first to take the ride. All seven wanted to be first. It was a test pilot's nature. No one wanted to play second fiddle, not in spaceflight or any other contest. When asked at their first press conference which one of them would be first, all seven raised their hands. But it wouldn't be up to them to make that determination, or to decide when the nation would launch a man into space.

The group was a good cross-section of military test pilots: Three from the Air Force, three from the Navy, and one Marine; ages 32 to 37; IQs ranging from 135 to 147; all married with children; and all eager to be first. There were those among them with egos, to be sure. It seemed to come with the territory. But of the seven, one would be hard-pressed to find a nicer fellow than Gus. He was the shortest of the Mercury Seven at just 5'7" and weighed about 155 pounds. His dark hair was always in a military-style crew-cut. Though he was by many accounts "the nicest guy," Gus always seemed to have a serious, determined look on his face, some might even say a scowl, as though he were an unhappy grump. Members of the media even took to calling him "Gloomy Gus" or the "Great Stone Face," he would later write. That was "because I tend to clam up at press conferences." Most of that, though, was a media-driven narrative. Those who knew Gus saw him as he really was. "I got to know Gus really well and he was just a really nice guy, straight-shooter, just like someone you might meet on the street. He never said a foul word about anybody," said John Boynton, who was a member of Chris Kraft's staff at the Manned Spacecraft Center in Houston. One who knew him as well as anyone was his younger brother Lowell. "A lot of people have a misconception about Gus," he said, "especially the media because he didn't care for the newspapers and any of that. A lot of people thought he was an introvert. I would say he was actually a mild extrovert. He was very outgoing with the right people."[26]

He certainly wasn't cranky by nature, but he could be a man of few words. When Gus flew cross-country with his equally quiet pal Deke Slayton, those trips, John Glenn later remarked, were "the least talkative flights…East Coast to West Coast in ten words or less." Once, during the Mercury program, Gus was asked to deliver a pep talk to the workers at the Convair plant, where the Atlas booster was being constructed, the very rocket that would take the Mercury capsule into orbit. His speech lasted all of a few seconds, "Do good work," he told them. And that was all. But that was Gus. His words, though, became a slogan of sorts for the workers at Convair, what one of Gus's biographers called "the shortest pep talk in history."[27]

Despite his quietness, his fellow astronauts, and others at NASA, came to like and admire Gus a great deal. "Grissom was good," said Flight Director Chris Kraft, noting that Gus had the reputation of being "a pilot's pilot, a talented engineer, and easy to like." Gus "was shy around people he didn't know," wrote Ed Buckbee, a member of von Braun's rocket team. "He was the smallest of the group…. Naturally tough and determined, he never thought of himself as a hero." Some of his friends and colleagues considered him "rough around the edges, but a good stick and rudder man." Jim Lovell called him a "pioneer" and "one of the great heroes of the space age." For Scott Carpenter, "Gus didn't say a lot, but when he did speak, it was worth listening to." Alan Shepard remembered being "impressed with Gus as an individual and as a pilot." He "didn't talk too much, but when he did...he really meant what was said. He was considered a great pilot, enthusiastic and competitive like the rest of us." Gordo Cooper was also close to Gus, "like brothers," he said. "We bitched and moaned to each other, patted the other on the back when he needed it, worked together on souping up race cars, and generally had a ball together."[28]

But when it came to work there was no one more intense than Gus, his low-key, quiet manner quickly disappeared, giving way to his focused work ethic. "Gus demanded the best," wrote his fellow Mercury astronauts Alan Shepard and Deke Slayton, "and he had an eagle eye for missed detail and a short fuse for slop. He could handle honest mistakes just as long as they didn't grow but when they did, his irritation took dead aim on the blockheads responsible, which he wanted off his team right then and there. Petty excuses were to him utter bullshit, and woe to the incompetent he found working on the ship his life depended on." For Gene Cernan, "Gus didn't mince his words or his actions. If Gus didn't like something, he let people know it." Apollo 7 astronaut Walt Cunningham, who was on the backup crew for Apollo 1, wrote that Gus was "a taciturn, grizzled fellow," who "filled the bill as the prototypical fighter pilot," who "could be cranky and tough," but he was also a "decisive guy, a team leader and an independent thinker, who nevertheless encouraged input from the rest of the crew."[29]

■　■　■

As America entered the 1960s, it was apparent that the Soviets possessed a great lead in the space race, despite aggressive American efforts to catch up. On November 8, 1960, America elected a new, vibrant president to succeed the aged Eisenhower. John Fitzgerald Kennedy, a U.S. senator from Massachusetts, became the youngest man ever elected to the nation's highest office. Throughout his 1960 campaign for the presidency, which he had been running for the most part for the previous four years, Kennedy consistently derided his opponent, Vice President Richard Nixon, and the Eisenhower administration, as old-fashioned and out of touch. "I don't mind sticking it to old Ike," he told his closest aides. And stick it to him he did, regardless of Ike's popularity. Eisenhower's presidency, JFK maintained, was outdated. Ike maintained a "detached, limited concept of the presidency," acting more like a "bookkeeper who feels that his work is done when the numbers on the balance sheet come out even." Rather than "ringing manifestoes issued from the rear of the battle," the next president had to "place himself in the very thick of the fight" with the Soviets. And that fight included space.[30]

"Our lagging space effort was symbolic, he thought, of everything of which he complained in the Eisenhower Administration: the lack of effort, the lack of initiative, the lack of imagination, vitality, and vision," said Theodore Sorensen, Kennedy's speechwriter, "and the more the Russians gained in space during the last few years in the fifties, the more he thought it showed up the Eisenhower Administration's lag in this area damaged the prestige of the United States abroad." Kennedy science adviser Jerome Wiesner had similar feelings. "I think he became convinced that space was the symbol of the twentieth century. He thought it was good for the country. Eisenhower, in his opinion, had underestimated the propaganda windfall space provided to the Soviets."[31]

The American press had similar feelings. The *New York Herald Tribune* published a series of articles on the big issues that would face the nation in the new decade. One covered the race to space and America's second-place standing in the world, a position that would remain unless the United States acted decisively. "Present indications are that unless we quickly and markedly accelerate our own efforts, the Soviets will be the first on the moon—and also, despite our dramatic Mercury astronaut program, the first man in orbit is quite likely to be a Russian. Whether we like it or not, we are in a race—a race that we cannot afford to lose," the editors wrote. "We can't win unless we start running. The free world has a right to demand more vigorous leadership than it has gotten from the United States in the competitive conquest of man's ultimate frontier." One columnist labeled the Ike years as a "time of great postponement." The *Washington Post*, in an essay entitled "The Fateful Decade," cautioned the nation about its secondary standing. "It is almost as if a deliberate decision has been taken to accept second class status. This continued slippage also affects starkly the challenges that lie ahead in the next year and decade," the paper opined. "Some of the lag in…space exploration may indeed be well-nigh irremediable. The deficiencies will place a heavy obligation upon the new President and Congress for the utmost candor and realism in acquainting the people with the hard facts as they relate to budgets and taxes."[32]

These same policy themes had been the central focus of Kennedy's aggressive attack on Nixon and the Eisenhower administration throughout 1960. During the campaign he remarked, "The first man-made satellite to orbit the earth was named Sputnik. The first living creature in space was Laika. The first rocket to the moon carried a red flag. The first photograph of the far side of the moon was made with a Soviet camera. If a man orbits the earth this year his name will be Ivan." Eisenhower had pushed back against the narrative that America was behind the Soviet Union, as did Nixon. By the fall of 1960, the United States had launched twenty-six satellites, while the Soviets had just six satellites. The Soviets had a slight edge with lunar probes, but America had launched the first interplanetary space probe, Pioneer 5.[33]

Kennedy, though, knew that none of that mattered one bit if the Soviets were first in putting a man in space. "If the Soviet Union was first in outer space, that is the most serious defeat the United States has suffered in many, many years," he said. "Because we failed to recognize the impact that being first in outer space would have, the impression began to move around the world that the Soviet Union was on the march, that it had definite goals, that it knew how to accomplish them, that it was moving and we were standing still. That is what we have to overcome, that psychological feeling in the world that the United States has reached maturity, that maybe our high noon has passed...and that we are going into the long, slow afternoon." A slight majority of the American people agreed with Kennedy, who won the election over Nixon by 112,000 votes out of more than 68 million cast.[34]

Although he had hammered the previous administration for sluggishness, by the time Kennedy was inaugurated president on January 20, 1961, Mercury was moving forward. As was his prerogative as president, Kennedy would name a new administrator for NASA, replacing T. Keith Glennan. Kennedy knew he didn't want a scientist as head of the space agency but someone who understood Washington, a man who could get things done. The president also knew that his vice president Lyndon Johnson "knew more about space than anybody," so he asked him to conduct a search for the right administrator. Johnson then asked Senator Robert Kerr of Oklahoma, a friend and his successor as chairman of the Senate Committee on Aeronautics and Space Science, if he had any suggestions. Kerr recommended James E. Webb, who had served as a pilot in the Marines in the 1930s and began his political career as director of the Budget Bureau, and later served as undersecretary of state under Harry Truman. At the time he was working as an executive at the Kerr-McGee Oil Company, owned by Senator Kerr. Johnson, who also knew Webb, interviewed twenty people but eventually came to believe that Webb was "the best possible man for the job" and sent his name to Kennedy.[35]

Webb, though, was reluctant to take the job. Needing advice, he went to his old boss, Senator Kerr. "Well, I think you ought to at least consider this," Kerr told him. According to Webb, there was no pressure from Kerr for him to take the job. But there was from Lyndon Johnson. "He was very anxious for me to do it," Webb said in 1969. LBJ "threw Frank Pace out of his office practically for suggesting that I was not the right man. I got the strong impression that he had come to the point that he wanted me to take this job. Now, whether that meant that he had tried so many others and not succeeded, or some other factor, I don't know but there wasn't any doubt in my mind that he wanted me to say yes that day." Johnson later wrote, "The choice of Webb as Administrator of NASA was one of the best selections I ever made. Under his forceful administration for eight years, America's space program progressed steadily from dream to fulfilment."[36]

True to form, Webb was no simple dreamer or stuffy academic, but an effective administrator who knew how the bureaucracy worked, which was exactly the kind of man Kennedy was looking for. Gemini and Apollo astronaut Frank Borman said "no finer public servant ever lived than Webb," a man who "was a skilled administrator with guts, vision, motivation, and dedication—the same qualities he would expect, in turn, from those working under him." Deke Slayton later wrote that "Webb was a real operator," who may not have been "technically trained, but he knew how to get things done. He turned out to be the best administrator NASA ever had." For Lyndon Johnson, Webb was more than that; he was the "best...administrator we have in this government." After six years at NASA, *Fortune* magazine wrote that "the evidence is persuasive that the nation has been well served" by Webb's leadership.[37]

■   ■   ■

After months of intense training, the decision of which American astronaut would be first into space came down to the NASA brass, specifically Robert Gilruth, who was the head of the Space Task Group, an assemblage of engineers within the agency that ran the Mercury program. He walked into the astronaut's office one day and made no bones about it: The first man in space would be Alan Shepard, followed by Gus Grissom, with John Glenn as back-up for both missions. Glenn would also get the third flight, which would likely be the first orbital mission. The selections made perfect sense from an aviation standpoint. Shepard was judged to be the most intelligent of the seven, so he would go first because of the unknown nature of the flight. Gus was the best engineer, so he would fly second to analyze everything and make sure it all worked correctly. Glenn was a superb pilot who could be counted on to return safely, no matter what happened, so he would be the best pilot to extend the mission to a new realm. But the decision had to be kept secret, especially from the press. And NASA took great strides to maintain the strictest confidentiality, right up to launch day, set for some time in early 1961.[38]

Shepard, a superb Navy flier, was now poised to become the first man in space in a suborbital flight, finally giving America something to brag about in the race with the Russians. A quick test with a chimpanzee named Ham on January 31 would set the stage for Shepard in March. But the chimp test was rife with problems. With Ham at the controls, instead of a man, malfunctions sent the capsule much higher and farther than NASA had planned, causing the chimp to pull an astonishing 16 Gs, twice what engineers had predicted. Instead of flying 300 miles downrange, Ham sailed 422 miles. And after slamming hard into the ocean, what emerged was a very unhappy, though healthy, chimp. Yet even though engineers quickly determined what caused the problem, a minor electrical relay switch, there was no way NASA was going to risk a man on the next flight. They needed to conduct one more test before Shepard would be allowed to fly, postponing his flight for weeks. The Mercury Seven were not happy. Would this give the Russians a chance to be first?

The answer to that question was a decisive yes. In the midst of the tests and delays, the unthinkable happened. On April 12, 1961, the Soviets once again stunned the world by putting the first man in space, twenty-seven-year-old cosmonaut Yuri Gagarin on his Vostok 1 mission. But more impressive still was that the Russians had managed to put their man into orbit with a rocket capable of pushing the spacecraft, which weighed nearly 10,500 pounds, up to 17,500 miles an hour. At that speed, Gagarin was able to make it into orbit by overcoming the gravitational pull of the planet. In the sixteenth century, Ferdinand Magellan first circumnavigated the globe in a wooden sailing ship that took three years to complete; Gagarin did it in an hour and a half. He then plunged back to earth after one revolution before parachuting from his capsule, officially negating the feat, as international flight rules stated that in order for a voyage to "count" in the books, a pilot had to land his craft safely. Gagarin had bailed out prematurely. But that hardly mattered to an astonished world, who wouldn't find out until years later that the Soviet had ditched. What mattered was the Russians had won again; they had reached space first; the rules be damned. America was bringing up the rear.[39]

That fact was made embarrassingly clear to the whole world with a very careless statement by a NASA spokesman. When contacted by a reporter in the middle of the night to get a reaction to news of Gagarin's flight, the groggy, half-asleep spokesman said, "We are all asleep down here." Headlines the next day said it all: "Russian in Orbit; America Still Asleep."[40]

Shepard was incensed upon hearing the news early the next morning. "You've got to be kidding me," he remarked when told of Gagarin's feat, his chance at immortality now forever vanquished. "We had 'em. We had 'em, and we gave it away."[41]

To the press, John Glenn laid it out more succinctly. "They just beat the pants off us, that's all. There's no use kidding ourselves about that," he said.

Five days later, America was humiliated yet again, though this time not in space, but with the disastrous Bay of Pigs fiasco in Cuba. Once again, the Soviets looked triumphant and strong while America looked weak and pathetic.

Somewhat good news wouldn't arrive until a couple of weeks later on May 5, when America finally answered the Soviets by sending Alan Shepard into space. But Shepard's feat was far less impressive than Gagarin's. A suborbital flight, Shepard's Mercury spacecraft, named *Freedom 7*, was boosted to an altitude of 115 miles into space, traveling 300 miles downrange and landing in the Atlantic Ocean, where naval recovery vessels were waiting. Total flight time: 15 minutes from launch to splashdown, with just five minutes of actual weightlessness. While the Soviets had managed to send their cosmonaut into orbit, the American Mercury Redstone rocket was incapable of pushing the spacecraft up to the appropriate speed to reach orbit, proving to the entire planet that America was lagging behind in rocketry.

Or was it? President Eisenhower's science advisor James R. Killian defiantly claimed that Soviet space feats were nothing more than efforts "to present spectacular accomplishments in space as an index of national strength." The Soviets knew full-well that they did not have the technological capabilities available to the United States, particularly in the long-term. So, many of the Soviet "feats" in space, and elsewhere, were gimmicks, making full use of "smoke and mirror" tactics, which would only continue as the space race ground on. Although their rockets were certainly more powerful than American ones at this early stage, and they appeared always to succeed on their first try, the tricks obscured the real state-of-play in the space race. The Soviets could easily mask failures and mistakes with their totalitarian system that did not allow for freedom of the press, or any other kind of freedom for that matter, despite their lofty rhetoric to the contrary. They tested their rockets and launched their missions from an area of their vast territory that was out of sight from much of their citizenry, at their space center in Kazakhstan, the Baikonur Cosmodrome. And with no press allowed, it was easy to hide disasters and showcase only successes, making it look as though they never failed.

Soviet archives, opened after the end of the Cold War, later revealed that they had failed almost as frequently as their American counterparts, with four failures in 1959 alone, while an ICBM explosion in October 1960 had killed 200 people. Unsurprisingly, none of these potential embarrassments made the headlines at *Pravda*. America, by contrast, with a free press and an open society, had to conduct her experiments in full view of the entire world.[42]

NASA believed in a slower, more methodical, but very hi-tech approach, taking the mission to land a man on the moon one step at a time. Gemini and Apollo astronaut Gene Cernan wrote that for the space program "everything was designed to advance in baby steps." Such an approach allowed for the utilization of American technology prowess. "This was the key element of the Mercury, Gemini, and Apollo projects—each succeeding mission was one more step toward the lunar landing goal," said Frank Borman, who also flew in both Gemini and Apollo. "Every flight proved something new; every step was deliberately cautious, to reduce risk as much as possible. Patience, in fact, was synonymous with safety, for each mission operated within its established parameters. As the unknowns were progressively exposed and understood, the parameters were extended."[43]

For the Soviets, haste was paramount, as well as power, for the sake of propaganda. Perhaps great in the short-term, their strategy would ultimately fail over the course of the decade. In one conspicuous example, Russia's rocket genius, Sergei Korolev, their version of von Braun, was considering how many cosmonauts his space program needed. First, he asked, "How many do the Americans have?" When told the number was seven, Korolev replied, "Give me three times as many." While the story may be apocryphal, it nonetheless provides solid insight into the Soviet mindset. They were out for quick political victories in the international press, not for the methodical approach that would be required to put a man on the moon.[44]

Though the Russians began developing big, powerful boosters immediately after the close of the Second World War, the United States wouldn't investigate such technology until much later. Contrary to popular belief at the time, this wasn't because the United States was technologically lacking—quite the opposite. The United States did not need the massive boosters the Soviets built in order to launch nuclear warheads anywhere in the world because the Americans had the ability to create smaller, less weighty bombs that still possessed the same explosive yield as a larger one. America had developed these "efficient" yet "still very powerful bombs," as Eisenhower had put it during a press conference in 1959. "This meant that we did not have the same power in our engines or in our boosters that was required if those warheads had not been so efficiently designed and built." It is a concept known as miniaturized warheads. This was also done because most of the warheads would be delivered by bombers, which could not carry an unlimited amount of weight.[45]

In addition, American guidance systems were much better and less bulky, which lightened the load. Khrushchev like to boast that Soviet missiles, which he claimed Russian factories were churning out "like sausages," had better guidance systems, a claim that turned out to be false. "It always sounded good to say in public speeches that we could hit a fly at any distance with our missiles," the blustering Premier once said, but "I exaggerated a little." Son Sergei was more truthful in his admission: "We threatened with missiles we didn't have." With America's use of superior technology, smaller missiles would do the same job as massive ones, and were much cheaper. The Russians always believed in the factor of numbers, size, and speed, not in putting together a superior technological product and the proper pace to make it happen successfully. "We concentrated on sophistication," Gus wrote, "while the Russians went the sheer-power route." The finish line of the moon race would show the flawed Soviet philosophy for all to see.[46]

Understanding the situation better than his older predecessor, President Kennedy saw the value of the space program and what it could do in the eyes of the world, and he wanted to make sure his nation did not fall further behind. Kennedy took a much more active approach in spending for both the military and NASA. But what would be the ultimate objective for the space race? How could the United States prevail in the contest with the Soviet Union?

To answer those questions, President Kennedy relied on Lyndon B. Johnson, his vice president. Kennedy knew what impressive work Johnson had done as a senator in the 1950s to spur investment in missiles, rocketry, and the budding space program. His hearings after Sputnik gained him national exposure. *Life* even put him on the cover under the title "Man of Urgency." So it was only natural that Kennedy turned to LBJ eight days after Yuri Gagarin's historic flight, when he sent a memo to Johnson seeking "an overall survey of where we stand in space." Specifically, Kennedy wanted to know what NASA's goals should be in relation to the Soviet lead in the space race—what does America need to do to beat the Soviets? A laboratory in earth orbit? A trip around the moon? Or a lunar landing? Furthermore, President Kennedy asked if the nation was "making maximum effort" and "achieving necessary results" in the effort thus far. JFK's science advisor, Dr. Jerome Wiesner, wanted him to push for unmanned projects, which would cost far less than manned missions and be a lot safer as well. But should Kennedy go for the less-risky option?[47] Only Johnson, well aware as he was of both the science and the political calculus, could help the young president answer these questions.

The greatest prize was obviously the moon, and landing a man there had already made its way into JFK's thinking before Gagarin's flight. At a White House meeting on March 22 attended by NASA Administrator Jim Webb, Deputy Administrator Hugh Dryden, Associate Administrator Dr. Robert Seamans, Atomic Energy Commissioner Glenn Seaborg, Vice President Johnson, National Security Advisor McGeorge Bundy, and JFK science advisor Jerome Wiesner, part of the discussion centered on restoring funding for the S-2 engine, which would power the Saturn V's second stage, the rocket system that could take men to the moon. Seamans told the president that funding the S-2 "would give us an option for preliminary Apollo flights in 1965, circumlunar flights in 1967, and a lunar landing in 1970 or soon thereafter." Kennedy agreed and approved the funding with an additional $70 million tacked onto it. Clearly, JFK had the moon in his sights.[48]

On April 13, when Seamans testified before the House committee overseeing NASA, the subject of a potential moon landing came up. One member asked if NASA thought the Russians would try for the moon by 1967, an important date for the Soviets, being the fiftieth anniversary of the Bolshevik Revolution of 1917, and if NASA could possibly get there by then. Seamans believed it was "achievable" if Congress supplied the proper funding.[49]

When the president asked for an assessment of the space race, Vice President Johnson, aided by Webb, von Braun, Secretary of Defense Robert McNamara, Secretary of State Dean Rusk, and "three distinguished Americans from the private sector"—Frank Stanton of CBS, Donald C. Cook of the American Electric Power Company, and George R. Brown of Brown and Root—sent Kennedy a six-page memo answering the president's every question. Acknowledging that "the Soviets are ahead of the United States in world prestige attained through impressive technological accomplishments in space," the Johnson memorandum admitted that America, with its "greater resources" had "failed to make the necessary hard decisions." The memo recommended moving forward "with a bold program." This would ensure the United States would be "the winner in the long run" by pursuing goals that the Soviets could not match. Johnson recommended a lunar landing mission as the best way to beat Russia in the great space game then underway. This could be accomplished by 1966 or 1967.[50]

Three days after Shepard's flight, Kennedy hosted America's first man in space, as well as the rest of the astronaut corps and NASA's top officials, at the White House for a ceremony, awarding Shepard NASA's Distinguished Service Medal in the Rose Garden. Afterward, the president wanted a briefing from NASA's top officials—Webb, Gilruth, Low, and others—on what the future held for the space agency. "The president was impressed with the world's reaction to the Shepard flight and wanted to know more about what we were going to do," Gilruth said.[51]

Gilruth had previously ordered a study of a moon program in mid-1960, but it did not call for a landing but a lunar fly-by mission. Gilruth, along with George Low, briefed Kennedy on what he could expect from the remainder of Mercury and also told him about the possible moon flight.

Kennedy, armed with information from the Johnson memo, then interjected. "Why aren't you considering landing men on the moon? If we're going to beat the USSR, don't we need to do something more than just flying around the moon?"

Gilruth was shocked by Kennedy's statement and assured the president that landing on the moon would be "an order-of-magnitude challenge than a circumlunar flight." But Kennedy didn't let go.

Seeing Kennedy's persistence, Gilruth told him that such a feat was possible. "What do you need?" Kennedy asked.

"Sufficient time, presidential support, and a congressional mandate," Gilruth answered. "How much time?" the president wanted to know. After talking it over with Low briefly, Gilruth replied, "Ten years."[52]

Less than three weeks later, JFK strode confidently into the House chamber with his usual confident swagger to a cheering throng of congressmen and senators. His intention on that day in May 1961 was not politics, even though he was less than a month removed from the Bay of Pigs disaster and changing the national subject might have been a wise political move. He was not there to rally his base, but to inspire the nation and to set it on a new course. The speech was entitled "Urgent National Needs" and touched on a number of subjects, but people only remember one: the promise to put a man on the moon.

President Kennedy's original draft called for landing on the moon by 1967, which Seamans had suggested was "achievable," as did LBJ's memo. But top NASA brass, including Seamans, "were aghast" at the proposal, which they had seen prior to the speech, leading Webb to phone Ted Sorensen, JFK's leading speechwriter, and later Kennedy himself, to push the goal back to at least the end of the decade. The president agreed. "I believe this nation should commit itself to achieving the goal, before this decade is out, of landing a man on the moon and returning him safely to the earth. No single space project in this period will be more impressive to mankind, or more important for the long-range exploration of space; and none will be so difficult or expensive to accomplish," the president said. "But in a very real sense, it will not be one man going to the Moon—if we make this judgment affirmatively, it will be an entire nation. For all of us must work to put him there."[53]

And lots of work it would take. If Kennedy's proposal shocked the assembled members in the chamber they didn't much show it, though a number of them were taken aback. But the top engineers at NASA, as well as the Mercury Seven, certainly were and were not hiding it.

"Did I hear what I think I heard?" Shepard asked aloud to his assembled colleagues at their office at Langley. Slayton nodded in agreement.

Gus, ever the wordsmith, simply snorted, "Kennedy is nuts" from his usual spot in the back of the room.[54]

Even though Webb knew what was coming, it was still shocking to actually hear the president say the words they had agreed to in front of Congress and the entire nation. Landing a man on the moon was now official policy. Gilruth, the longtime aerospace engineer and former pilot, was as shocked as he had been in the Oval Office. "I could hardly believe my ears," he said. "Frankly, I was literally aghast at the size of the project.... I wasn't at all sure it could be done." Gilruth was particularly concerned with the word "safely," to "return him safely to the earth."

"That's not simple. We don't know how to do that," Gilruth said. When Flight Director Chris Kraft heard the speech, he was, at least for a moment, "paralyzed with shock," he later wrote. "My mind was going off in a hundred directions." Another NASA flight director, Glynn Lunney, was equally shocked. "When I first heard President Kennedy's speech...I was overwhelmed at the magnitude of it. I mean, we were struggling with a Mercury spacecraft that weighed 2,000 or 2,500 pounds. [Now] we were talking about a spacecraft that would be 10 or 20 times bigger. For me it was just an overwhelming thought that we could actually go and land on the moon and bring somebody back."[55]

Former President Eisenhower, like Gus, thought the idea was "just nuts," but while Gus was considering the technology that would be needed but was yet to be invented, Ike was thinking in terms of cost, especially if the plan called for the country to spend as much as $40 billion to do it. But President Kennedy wasn't so conservative as to worry about who would foot the bill. He believed in America, that it could hit the lofty target. And he knew the Russians couldn't. What's more, the Russians knew they couldn't, at least not in that time frame. So, now that the objective was out there, NASA rolled up its sleeves and went to work 'round-the-clock to make it happen. Kennedy did his part and pressed Congress for funds. A few months later, in September, he signed a special appropriation to jump-start the Apollo program.[56]

■   ■   ■

Gus had more on his mind than Kennedy's ambitious plan. He was soon to be the second American in space, following the same path as Shepard on another Mercury-Redstone suborbital. He would travel to an altitude of 115 miles, arching 300 miles downrange to a splashdown in the Atlantic. Scheduled to launch in July, Gus began working hard on the mission months before Shepard's flight, first laying eyes on his capsule in January at the McDonnell plant in St. Louis. Gus worked closely with the engineers at McDonnell, as well as management, attending as many meetings as he could. "I thought it would be good for the engineers and workmen who were building my spacecraft to see the pilot who would have to fly it hanging around. It might make them just a little more careful than they already were and a little more eager to get the work done on time if they saw how much I cared." It was a philosophy he maintained throughout his astronaut career.[57]

His spacecraft, *Capsule 11*, was sent to the Cape on March 7 for continued work. Over a period of thirty-three days, the craft would be disassembled into its component parts, each part re-tested, then re-assembled, and tested as an entire system once again. "We kept spotting problems," Gus wrote, "as we knew we would. But there were very few of them, considering the state of the art, and the simulations we went through for practice went very well." But, Gus being Gus, he did get "slightly impatient whenever some technician came up with a new modification in the system that might have caused a long delay if we had accepted it."[58]

The craft was an improvement over the one Shepard had used. It had a window so Gus could see out into space and view the earth; Shepard had only a periscope. Gus's capsule also had a new hatch. The old hatch had a latching mechanism but the new one had explosive bolts, an improvement which was sorely needed for later orbital missions. The new explosive hatch had been added to the spacecraft design in response to the Mercury test flight with Ham, when the capsule began to sink before it was snatched up by the attending ships. Astronauts needed a way to get out of the capsule fast should the same thing happen on a manned flight, and it also allowed rescue crews to get into the spacecraft faster if they needed to after splashdown. Unbeknownst to anyone, testing the new hatch would be the central focus of Gus's mission.[59]

Gus was raring to go and wanted no needless delays or unnecessary risks. "I gave up water skiing and I was more careful than usual to observe the speed limits around the Cape," so as not to do anything that might cost him his mission. Gus and his fellow pilots were known to race around the area in their corvettes, oblivious to speed limit signs or ordinary rules of the road. Now that had to end, at least until after splashdown.[60]

On July 21, 1961, aboard his *Liberty Bell 7* spacecraft, Gus roared into the heavens on top of a Redstone rocket, pushing through the clouds and into the blackness of space. Reaching the apogee of his flight, officially 118.2 miles, he looked out his window, which had not been available on Shepard's flight, and could clearly see the Cape and other distinct features around it, like the Banana River and even individual buildings. Everything went perfectly, until splashdown that is. "I hit the water with a good bump," he later wrote. "The capsule nosed over in the water, and the window went clear under. Almost immediately I could hear a disconcerting gurgling noise. But I made a quick check and could see no sign of water leaking in." The spacecraft eventually righted itself to an upright position after about thirty seconds. "I felt like everything was in very good shape." He soon began to go through his checklist of assigned tasks while waiting on the helicopter to recover him. "I had opened up the faceplate on my helmet, disconnected the oxygen hose from the helmet," then began releasing the numerous straps that kept him in place. He then did something that would become very important in the moments ahead. "I rolled up the neck dam of my suit, a sort of turtle-neck diaphragm made out of rubber which we tighten around the neck when we take the helmet off to keep the air inside our suit and the water out in case we get dunked during the recovery. This was the best thing I did all day," he wrote.[61]

While bobbing in the ocean waiting on the Navy to pluck him from the sea, the new hatch on Gus's capsule, engineered with explosive bolts to allow for an easy escape if necessary, prematurely blew. The hatch flew away from the capsule and water began pouring in.

Confusion soon hit Mission Control. The radio link connecting flight controllers with the spacecraft was garbled, so much of the exchange between Gus and the helicopter was indecipherable. Questions began flying between controllers and the flight director, Chris Kraft. "Is he in the water?" "Do they have the capsule hooked up?" No one could really say for sure.[62]

After the mission Gus told investigators exactly what happened from his vantage point. "I was lying there, minding my own business when, POW, the hatch just went. And I looked up and saw nothing but blue sky and water starting to come in over the sill." Gus had gone through his checklist in regards to the hatch. He removed the cap from the detonator and pulled out the safety pin, thereby arming the detonator but "I did not touch it," he insisted. Procedure called for him to blow the hatch once the chopper had a firm grip on the capsule, so he waited to hear from the pilot to give his okay to blow the hatch.[63]

When the hatch blew on its own, Gus had to bail out quickly, floundering in the ocean and fighting swift Atlantic currents as well as the downdrafts from the chopper swirling above. The neck dam of his suit prevented water from pouring in, but a hose valve on the front of his suit was open and water began seeping in through that opening. Luckily, Gus was able to get his suit closed during his struggle for survival. The Navy chopper, prioritizing recovery of the spacecraft, was unable to hoist the capsule because of the weight, which increased to five thousand pounds due to the mass of water now inside, at least one thousand pounds above the lifting capacity of the helicopter. After releasing it, the capsule sank to the bottom of the Atlantic Ocean.[64]

Bewilderment still pervaded Mission Control. "Oh, hell, I think it sank," said one controller. "We didn't know what happened to Gus," Kraft said. "After a minute of silence, a second helicopter reported that they were 'attempting to recover the astronaut.'"

"What does that mean?" one controller asked. But Kraft finally snapped at all the chatter. "Shut up! Shut up and listen!"

Were they awaiting the worst? "I remember thinking, *I hope it's not a body they're recovering*," Kraft later wrote. "The next minute dragged on forever. Then we heard it. Gus was aboard the chopper. They were returning to the ship."[65]

Gus was safe but completely exhausted after nearly drowning—the amount of water in his spacesuit had threatened to pull him under. The chopper crews "did not realize how much trouble I was in," Gus noted later. But Gus also did not realize how much trouble he was in. Just before bailing out of the spacecraft, he ditched his helmet. While on the carrier deck, an officer brought it to him, having recovered the helmet after it floated out of the capsule. "For your information," the officer told him, "we found it floating right next to a ten-foot shark."[66]

Shepard's flight was met with wild celebrations, a ticker-tape parade, and a trip to the White House to meet with President Kennedy, where he was awarded a medal in the Rose Garden. Gus was not so fortunate. He faced mounting questions and numerous inquiries both inside and outside NASA. Mercury astronaut Scott Carpenter said Gus "thought he'd been left to twist in the wind," facing the inquiry alone and with little support. Did the equipment malfunction? Or was it an accident? Did Gus panic and hit the button to blow the hatch? Gus maintained that he had done no such thing, that the hatch blew on its own, a defect in the manufacturing, not pilot error. Engineers were skeptical, citing test after test in which not a single hatch with explosive bolts had ever blown on its own. Walt Williams, a NASA engineer and director of operations for Mercury, did not think it was Gus's fault. "I never believed that he panicked and blew the hatch. I do believe, however, that it's very possible he bumped the switch by accident with the helmet." Although that was a possibility, which Gus later admitted to Slayton, it is unlikely to have occurred in such a manner.[67]

Gus was embarrassed and many thought his space career might be over. "It was especially hard for me, as a professional pilot. In all my years of flying, including combat in Korea, this was the first time that my aircraft and I had not come back together. In my entire career as a pilot, *Liberty Bell* was the first thing I had ever lost," he said. To make matters worse, after choosing the name for his craft, Gus had the *Liberty Bell* painted on the side of the capsule, complete with the authentic crack. "We got one joke out of it, though," Slayton wrote. "Never again would we launch a manned spacecraft with a crack in it." Jokes aside, Gus's days as an astronaut were not over, not by a long shot, for he was one of the best pilots and engineers in the astronaut corp.[68]

Gus knew he didn't do anything to prematurely blow the hatch. And Gordo Cooper believed him too. "From my experience flying with him, I knew Gus was a great pilot. When he returned from his space mission and insisted that he hadn't blown the hatch early…I believed him. I knew that if he had screwed up, Gus would have been the first to admit it."[69]

But he somehow had to prove it. For weeks after his mission, he sat inside other capsules and "tried to duplicate all of my movements, to see if I could make the same thing happen again." But he could not. "The plunger that detonates the bolts is so far out of the way that I would have had to reach for it on purpose to hit it and this I did not do," he said. "Even when I thrashed about with my elbows, I could not bump it accidentally."[70]

Wally Schirra, as a fellow astronaut, sided with Gus. He knew that there "was only a very remote possibility that the plunger could have been actuated inadvertently by the pilot." It was a "travesty that the movie *The Right Stuff* showed him as a babbling coward in a spacecraft," Schirra later wrote of his friend, for he was "quite convinced" that Gus had most certainly not blown the hatch. So Schirra decided to conduct a test of his own. After his own Mercury mission in the fall of 1962, while still inside the capsule on the deck of the recovery ship, Schirra hit the button to blow the hatch. The "recoil cut through my glove into my hand and made a big cut." Gus had no such injury. After the flight, Gus was on the carrier deck to greet Schirra. "I held up my hand and said, 'Gus, look at the cut I have on my hand. You didn't have anything like that at all on you did you?' The biggest smile I ever saw on the Pacific Ocean was on Gus's face which vindicated the fact if he had hit that button inadvertently, he would've had a real welt on whatever part of his body hit it." In fact, every Mercury astronaut who manually blew the hatch suffered some kind of injury to his hand, except Gus. So he could not have blown the hatch himself.[71]

Slayton agreed with Schirra's assessment, especially about *The Right Stuff*. "Everybody later developed an opinion about what had really happened. There was one group that thought Gus had screwed up...that he had blown the hatch early by mistake. This, of course, is my gripe with Tom Wolfe's *The Right Stuff*. He was kind of tough on Gus anyway, and in the matter of the hatch pretty much convicted him." Essentially speaking for the rest of the astronaut corps, Slayton defended Gus to the hilt. "We don't feel that he lost his spacecraft or that he goofed," he said in an interview in 1989. "We think there is pretty firm proof that he did not blow the hatch. Certainly, he would not have been given the first Gemini spacecraft and Apollo spacecraft if he really had goofed." And Slayton should know, since it was he, as the director of Flight Crew Operations, who selected every Gemini and Apollo crew.[72]

For Gus, "It remained a mystery how that hatch blew. And I am afraid it always will. It was just one of those things." In 1999, an exploration team found the *Liberty Bell 7* capsule on the ocean floor in sixteen thousand feet of water, three thousand feet deeper than the *Titanic*, and pulled it to the surface. But the hatch was never found. The spacecraft was restored and is currently in a museum in Kansas. Inspection of the craft, without the hatch, could not prove what happened. Writing in 2005, Schirra was adamant that his test alone should put "that ridiculous story" to rest once and for all time. "Gus did not do that and a movie that would portray Gus as that kind of man should be banned."[73]

■   ■   ■

Controversy aside, there was little time to worry over it. Only a couple of weeks after Gus splashed down, the Soviets flew their second space mission, Vostok 2, with Gherman Titov, a Soviet cosmonaut who remained in orbit for a full day—twenty-five hours, eighteen minutes to be exact—completing nearly eighteen orbits of the earth. Embarrassingly enough, America had yet to orbit the planet a single time, while the Russians were staying up for a day at a time. While the original plan called for six suborbital flights, NASA now knew the timetable had to change.[74]

One problem was a booster. The only working rocket in the U.S. arsenal at the time capable of boosting the capsule into orbit was the Atlas. But it had what can only be described as a serious defect: It tended to blow up more often than not. Slayton was the astronaut assigned to work on the booster problems, and he spent a considerable amount of his time "watching Atlases take off and blow up." The Atlas missile never had a problem when carrying a nuclear warhead, but mounting a 2500-pound spacecraft would cause the rocket to come apart when it hit Max Q, the area of maximum dynamic pressure on the booster. The problem turned out to be the missile's thin skin. When the skin was thickened, the Atlas's performance drastically improved.[75]

But as with Shepard, before NASA would man-rate the spacecraft and its launch system, it wanted another flight with a chimp. Enos, like his companion Ham, would soar into space in a Mercury capsule, but Enos would ride the Atlas into orbit. Enos launched on the morning of November 29, 1961, and was scheduled for three orbits but the mission was cut short after two revolutions because of problems with the spacecraft. During the flight, President Kennedy was holding one of his many press conferences, which he always handled with great poise and skill. Handed a note in the midst of fielding questions, JFK, using his quick wit, made light of the moment. "This chimpanzee who was flying in space, took off at 10:08. He reports that everything is perfect and working well." The assembled press broke out in loud laughter. But it was not perfect and working well. And the issues NASA was reporting would be no laughing matter with a man on board. When all the kinks were finally worked out, NASA felt confident enough to send the first American into orbit, John Glenn in February 1962, nearly a year after the Soviets.

On *Friendship 7*, Glenn was set to make at least three orbits of the earth. During his flight, he reported being surrounded by "brilliantly-lighted" particles which resembled fireflies. In a later mission, astronauts determined that the particles were not, in fact, creatures, but expelled gases from the spacecraft, which froze upon hitting the icy-cold void of space. But Glenn's time in space wasn't just looking out the hatch at the "fireflies"; matters turned much more serious on the third orbit. Instrumentation soon revealed that the capsule's landing bag had deployed while Glenn was still in space, which meant the heat shield might come off during re-entry. Held on by straps, the retro pack was essential for re-entry. After firing the retro rockets to slow down the spacecraft in order to re-enter the earth's atmosphere, the pack was jettisoned, exposing only the heat shield, which protected the craft from the intense heat of re-entry. The flight director, Chris Kraft, made the decision to keep the retro pack on, hoping it would keep the heat shield in place. The plan worked, and Glenn splashed down safely, becoming a major American hero. Although he would not fly in space again until 1999 aboard the Space Shuttle, he won a U.S. Senate seat in Ohio in 1974, which he held for a quarter of a century.

Scotty Carpenter, who had been Glenn's backup on *Friendship 7*, would eventually get the nod to try and repeat Glenn's feat. On May 24, 1962, Carpenter flew into orbit aboard his Mercury spacecraft, dubbed the *Aurora 7*. But Carpenter had troubles of his own, some said of his own making, while others contended it was difficulties within the spacecraft itself, minor flaws in the control system experienced on other

Mercury missions. NASA brass accused Carpenter of being distracted, worrying too much about the scientific experiments and not enough time spent on actual flying. Slayton called the flight "sloppy." For whatever reason, Carpenter was a few seconds late firing his retro rockets, which, at such a high rate of speed, put him 250 miles off course when he splashed down. It took rescuers longer to find him, drifting along in a life raft seemingly without a care in the world. Angry NASA officials, though, made sure he never flew in space again.

In September 1962, about a month before the next scheduled Mercury flight, President Kennedy traveled to Houston, Texas, the city chosen as the site of the newly planned Manned Spacecraft Center, the official headquarters of the astronaut program and the spot where controllers would monitor all flights beginning in 1965 from Mission Control. Speaking at Rice University, Kennedy, buoyed by two recent successful orbiting missions, laid out the vision he had for sailing "on this new sea," spelling out why America must go to the moon. It was one of his most memorable speeches, one in which he intended to rally the troops and galvanize the entire nation. "We choose to go to the moon," the president exclaimed. "We choose to go to the moon in this decade and do the other things, not because they are easy but because they are hard...because that challenge is one that we are willing to accept, one we are unwilling to postpone, and one which we intend to win!" For Kennedy knew full-well that the Soviets had their problems too. "We have had our failures, but so have others, even if they do not admit them. And they may be less public," he said. He admitted that the United States was "behind, and will be behind for some time in manned flight. But we do not intend to stay behind, and in this decade, we shall make up and move ahead." For "no nation which expects to be the leader of other nations can expect to stay behind in the race for space," he said. Kennedy had complete confidence that Americans could do great things and would land on the moon and return safely to the earth by the end of 1969.[76]

Being behind, though, was an understatement. The previous month, the Soviets again increased their lead by placing 2 spacecraft in orbit simultaneously. They did not rendezvous, a critical element for a lunar flight, coming within just 3.2 miles of one another, but each craft remained in orbit for lengthy periods of time with one, piloted by Andriyan Nikolayev, staying in orbit for nearly 4 days and circumnavigating the globe 64 times. NASA had managed but 3 orbits only twice, and both of those had been a struggle. And with these feats, scientists and those in the press praised the Russians and looked down at lowly America. After the dual spaceflight, one London paper headlined: "Two Upsmanship." Sir Bernard Lovell stated that it was his opinion that "the Russians are so far ahead in the technique of rocketry that the possibility of America catching up in the next decade is remote." Dr. Edward Teller, the inventor of the hydrogen bomb, believed the Soviets had a decided advantage. "There is no doubt that the best scientists as of this moment are not in the U.S., but in Moscow."[77]

In October, in a mission that came just before the start of the Cuban Missile Crisis, Wally Schirra would attempt to double the three-orbit flight of Glenn and Carpenter with his *Sigma 7* capsule. The mission was a complete success, with six orbits, though he did experience some of the same technical difficulties that Carpenter had on his mission. But even a nine-hour, six-orbit mission was nothing compared to what the Russians were doing.

In May 1963, Gordo Cooper flew the sixth and final Mercury mission and by far the most ambitious to date, hoping to at least get close to matching the Russians. In his Mercury capsule, which he dubbed *Faith 7*, much to the chagrin of NASA, Cooper pushed the spacecraft to its limits, remaining in space for more than a day and orbiting the earth twenty-two times. He was the first American astronaut to sleep and eat in space and the last to travel into the heavens alone.

The finale of the Mercury-Atlas 9 mission, though, would prove the most exciting yet. Cooper's machine began to shut down while still in orbit. Yet through it all he was never rattled. In a profession full of cool, calm, and collected test pilots, one would be hard pressed to find one cooler, calmer, and more collected than Gordo Cooper, so relaxed, in fact, that while sitting on the launch pad waiting for liftoff of his Mercury-Atlas he actually fell asleep for a quick nap. As things began to deteriorate on his mission, he unexcitedly relayed the distressful message back to the ground. "Well, things are beginning to stack up a little," he said in his conspicuous, slow Oklahoma drawl. "ASCS inverter is acting up. And my CO_2 is building up in the suit. Partial pressure of O_2 is decreasing in the cabin. Standby inverter won't come on line. Other than that, things are fine." A normal person would likely have panicked, but not a seasoned test pilot who was accustomed to taking such risks on an almost routine basis. That's why it was so crucial to utilize military test pilots on such daring missions.[78]

As the time approached to re-enter the earth's atmosphere, *Faith 7* was completely dead. All the automatic controls had quit and Cooper had to do everything manually. He had to look out the capsule's window, lock his eyes on the horizon, use his watch to time when to fire his re-entry thrusters, then manually steer the craft in its long plunge to splash-down, maintaining the proper attitude alignment so he would not burn up in the ever-thickening air. And he was almost dead perfect, landing closer to the rescue carrier than any previous flight.

■   ■   ■

President Kennedy was overjoyed with Cooper's flight and hosted the Mercury Seven for a ceremony on May 21, 1963. Awarding Cooper NASA's Distinguished Service Medal, JFK remarked how extraordinary the flight was, coming as it did on the anniversary of Charles Lindbergh's flight across the Atlantic. "I think one of the things which warmed us the most during this flight was the realization that however extraordinary computers may be, that we are still ahead of them and that man is the most extraordinary computer of all. His judgement, his nerve, and the lessons he can learn from experience still make him unique, and therefore makes man flight necessary and not merely that of satellites.

"I hope that we will be encouraged to continue with this program," Kennedy continued. "I know a good many people say 'why go to the moon.' Just as many people said to Lindbergh 'why go to Paris.'...I think the United States has committed itself to this great adventure in the '60s and I think before the end of the '60s we will see a man on the moon, an American."[79]

Later that year, in October, Kennedy invited all seven Mercury astronauts to the White House once again to present them with the Collier Trophy, an annual award established in 1911 and given for "the greatest achievement in aeronautics in America." It was named for Robert J. Collier of *Collier's Weekly* magazine, who promoted the cause of aviation generally and supported the achievements of the Wright brothers. It was one of the most sought-after awards in all of aviation. "Some of us may dimly perceive where we are going and may not feel this is of the greatest prestige to us. I am confident that its significance, its uses and benefits will become as obvious as the Sputnik satellite is to us, as the airplane is to us," Kennedy said at the ceremony. "I hope this award, which in effect closes out the particular phase of the program, will be a stimulus to them and to other astronauts who will carry our flag to the moon and perhaps someday, beyond."[80]

Kennedy admitted once privately, in a taped conversation with NASA officials, that he was "not that interested in space," meaning the nuts and bolts of the program. When Jim Webb came by the Oval Office one day to show the president a model of a future space capsule for human spaceflight, Kennedy was unimpressed, "I think he stopped by a toy store on his way to town," he later told Ted Sorensen. But Sorensen understood what was really going on in Kennedy's mind as it related to the space program, "It seems to me that he thought of space primarily in symbolic terms. By that I mean he had comparatively little interest in the substantive gains to be made from this kind of scientific inquiry. He did not care as much about new breakthroughs in space medicine or planetary exploration as he did new breakthroughs in rocket thrust or humans in orbit."[81]

JFK certainly knew what it meant for the country and the world. Although he wasn't interested in details, he focused on the big picture, and he wanted to get that across to Webb and the rest of the NASA team. "Everything you do ought to be tied to getting onto the Moon ahead of the Russians," he said. "The Soviets have made it a test of the system." And he wanted America, and capitalism, to prevail.[82]

And it seemed to be always on his mind, even the day before he died. Speaking in San Antonio, Texas, on November 21, 1963, six months to the day after he honored Cooper in the Rose Garden, JFK used a story from his family's ancestral homeland to continue to push his goal for the American space program.

> Frank O'Connor, the Irish writer, tells in one of his books how, as a boy, he and his friends would make their way across the countryside, and when they came to an orchard wall that seemed too high and too doubtful to try and too difficult to permit their voyage to continue, they took off their hats and tossed them over the wall—and then they had no choice but to follow them.
>
> This Nation has tossed its cap over the wall of space, and we have no choice but to follow it. Whatever the difficulties, they will be overcome. Whatever the hazards, they must be guarded against. With the vital…help of all those who labor in the space endeavor, with the help and support of all Americans, we will climb this wall with safety and with speed—and we shall then explore the wonders on the other side.[83]

Sadly, though, Gordo Cooper's final flight in the Mercury program would be the last one John F. Kennedy would live to see. Tragically gunned down on November 22, 1963, his dream of an American on the moon would have to continue under the new president, Lyndon Baines Johnson.

Despite a few snags, Project Mercury was a complete success. It was now time for the next phase, Project Gemini. The two-manned missions would further the goals needed to travel to the moon. New groups of astronauts would join the growing team, but just when NASA was gearing up for a repeat of the success of Mercury, with Alan Shepard leading the way, fate intervened, placing Gus Grissom on the top of the astronaut hierarchy. It would be Gus who would become the first man to travel into space twice.

II

THE BRIDGE: GEMINI

Ed White knew what he wanted to do with his career as an Air Force pilot. It was the late 1950s, only a few years since he had graduated from West Point and achieved his dream of becoming an aviator. He had recently read an article about the future of flying in space with pilots called astronauts. "The article was written with tongue in cheek, but something told me: this is it—this is the type of thing you're cut out for," he later said. "From then on everything I did seemed to be preparing me for space flight." America was moving into space, and Ed White wanted to be a part of it.[1]

Stationed in Germany at the time, Ed flew fighters along with his good buddy and future astronaut and moonwalker Buzz Aldrin. After a fun-filled day flying simulated combat missions over the North Sea with a new crop of rookie fighter pilots on a training hop, Ed and Buzz spent a few minutes talking as they waited for each pilot to land safely back at the base. As they stood next to their jets, Ed told Buzz that he planned to use an Air Force program that allowed qualified officers to attend graduate school. He planned to head to the University of Michigan and obtain a master's degree in aeronautical engineering. After

Michigan, he would enroll in the one-year Air Force Test Pilot School at Edwards Air Force Base in California. He had a purpose in all this. His plan, he told Buzz, was to become an astronaut and take his flying career into outer space.[2]

NASA had already announced Project Mercury by then and was in the process of choosing astronauts for the initial program. So, even though mankind had yet to reach space, Ed knew that spaceflight would soon be a reality. But he was not then qualified to apply for the new astronaut corps, lacking the education and test pilot requirements. Still, he knew that America would likely need more astronauts in the future, so he planned to be ready to accept the invitation to join the most exclusive flying club in the world when the call came. And he wouldn't have to wait long for that call to come.

In 1962, NASA wanted more astronauts. At this point, NASA had but four manned missions under its belt, but was already planning for the next program. The call for new astronauts went out in April that year with a deadline of June 1 for applications. The requirements were roughly the same as Mercury except that prospective astronauts could be as tall as six feet and would not be limited to military test pilots this time around; civilian test pilots could also apply. By the deadline, NASA had received 253 applications. They put the prospective candidates through the same litany of physical and psychological examinations as the Mercury applicants. John Young called them "tortures," a "series of medical and psychological tests, some of them pretty crazy." For example, one psychiatrist asked him, "Who do you hate more, your mom or your dad?" After the battery of extreme testing, NASA made their selections. And of the nine they chose, Ed White was among them, a group dubbed "The New Nine."[3]

Along with Ed, then an Air Force Captain, the New Nine were Neil Armstrong, a civilian test pilot for NASA; Air Force Major Frank Borman; Navy Lieutenant Charles "Pete" Conrad; Navy Lieutenant Commander James Lovell; Air Force Captain James McDivitt; Elliott See, a civilian test pilot for General Electric; Air Force Captain Thomas

Stafford; and Navy Lieutenant John Young. Although Armstrong and See, who became good friends, were listed as "civilian" test pilots, both had previous military service in the Navy, while Armstrong had combat experience in Korea on his resume as well. The New Nine were introduced to the world at a packed press conference at the University of Houston on September 17, 1962.[4]

It was a very talented group of flyers, six of whom would make lunar trips in Apollo—Armstrong, Conrad, and Young walked on the lunar surface, while Borman, Lovell, and Stafford made orbital flights to the moon. Lovell and Young flew to the moon twice. Deke Slayton, director of Flight Crew Operations, called the New Nine "probably the best all-around group ever put together." For Wally Schirra, the class was "superb," a "championship team." Gene Cernan, who would join NASA as a member of the third group of astronauts, said the New Nine were "as profound and talented a group of pilots I have ever seen, all top-notch aviators." Frank Borman wrote that he had "never known a finer group of men." Yet as far as spaceflight was concerned, they were all green. Slayton assigned Gus to supervise the new guys, and he couldn't help but be impressed with them. "They're all talented. In fact, when one of them comes up with a new answer for some problem, I think they are smarter than our group of seven," Gus said. The average IQ of the New Nine was 135. But Gus was also quick to put them in their place. "Don't feel so smart," he told them one day. "You're just an astronaut trainee. You're not an astronaut until you *fly*."[5]

The New Nine, though, wouldn't fly in Mercury, which, by the time they were chosen, was nearly over, but would prepare to begin a brand-new project. Originally, NASA had conceived of just two programs, Mercury and Apollo, but it didn't take long for engineers and planners to realize there was a lot more involved in flying to the moon than they had initially believed. New skills and procedures were needed, and it would take additional missions to acquire them. Mercury was simply not up to the job because it was not designed to do more than its original intention. It was very limited in scope, intended to put man in space and orbit the earth, nothing more. Apollo would be focused on getting to the moon. So aeronautical experts agreed that a new program would have to be conceived to act as a bridge between Mercury and Apollo for the sole purpose of acquiring necessary skills and techniques—long duration missions of up to two weeks in zero gravity in order to find out if a journey to the moon was even physically possible for a human being, master rendezvous and docking maneuvers that were far more complex than one might think, and conquer the fine art of spacewalking, officially known as Extra Vehicular Activity, or EVA. The "whole reason for Gemini was to lay stepping stones toward Apollo," wrote astronaut Cernan. For Buzz Aldrin, "Gemini was the proving ground for Apollo." While Slayton noted, "Most of the major unknowns relative to getting into the Apollo program came out of the Gemini program." If Gemini failed, there would be no Apollo. So the program was more than necessary.[6]

■   ■   ■

On December 7, 1961, with just two Mercury sub-orbitals in the books, Bob Gilruth, the man who ran the Mercury program as head of the Space Task Group, announced that NASA was planning a new follow-up program consisting of a two-man spacecraft to gain those necessary skills. Originally called Mercury Mark II, it was soon renamed Project Gemini, with an initial budget for the boosters, twelve capsules, and everything else needed to carry out a project estimated at half a billion dollars. At the time the full program was publicized, though, hardly anyone noticed, coming as it did during the national euphoria over John Glenn's orbital flight in February 1962.[7]

Gemini is a two-star constellation in the northern sky. The name "Gemini" is Latin for "twins." It fit perfectly with NASA's new program. Instead of one astronaut in the spacecraft, like Mercury, there would be two, because "the complexities of rendezvous and the strain of long duration flight" would require more than one crew member. The craft itself, built by the same manufacturer, McDonnell Aircraft Company, resembled the Mercury capsule in many aesthetic ways, but not in the way it flew. And Gus had a big role to play in spacecraft development. He was fully on-board from the very beginning.[8]

"We put Gus into the Gemini program—as the crew interface manager—very early," Slayton said. Soon after NASA announced Gemini, Gus began spending more and more time at the McDonnell plant in St. Louis where the craft was under construction. For him, it was like a re-birth. After the trouble with *Liberty Bell 7*, Gus "felt reasonably certain, as the [Gemini] program was planned, that I wouldn't have a second space flight," he later wrote. "By then Gemini was in the works, and I realized that if I were going to fly in space again, this was my opportunity, so I sort of drifted unobtrusively into taking more and more part in Gemini. What it amounted to, in fact, was that they just couldn't get rid of me, so they finally gave up and programmed me into Project Gemini."[9]

Perhaps he was being a bit modest. Gus was placed there because of his incredible skills as both a pilot and an engineer, working with techni-cians and designers to make Gemini much more flight-worthy than Mercury ever was. The "primary objective of Gemini was to advance the state of the art," Gus wrote. He defined the phrase this way: "Where we stand at the moment, what we know can be done because it has been done, and what we're pretty sure can be done in the future on the basis of the first two."[10]

It was easy for Gus and the rest of the astronaut corps to advance the state of the art by making recommendations, changes, and innova-tions to their spacecraft, since they possessed a wealth of aviation knowledge, as well as the fact that they were the brave souls who would actually fly it. Unlike Russian cosmonauts, American astronauts were "deeply involved in every step of the development of our spacecraft and their supporting systems," Gus said. "If one of us didn't like the solution the designers came up with, we said so without hesitation. If we thought we had a better idea, we were free to say so, and often have." And the engineers at McDonnell would listen, not taking the slightest offense at suggestions by the astronauts because of their vast experience flying untested machines and their advanced studies in aeronautical engineering. They knew what they were talking about. "I would sit in the mockup for hours," Gus would write in *Life* magazine. "All I had to do was say 'No, I don't like it' or 'Yeah, it's okay.' When the other [astronauts] started looking at the Gemini mock-up it was pretty clear it was designed around me."[11]

His fellow astronauts admitted as much. "We nicknamed Gemini the 'Gusmobile' because he was so much a part of that spacecraft," Slayton said. "We ended up with a machine that was really a manned machine in Gemini, one that could be piloted." Fellow astronaut Schirra also knew that it truly was the Gusmobile, "essentially his spacecraft. He practically had it to himself." As John Young, who would fly with Gus on Gemini 3, noted, "Gus really had a big hand in everything, from the way the cockpit was laid out to what instruments went where. It was his baby." The spacecraft was, in fact, so closely designed around Gus and his short frame that fourteen of the sixteen astronauts could not fit inside the original model, so additional changes had to be made in order to accommodate them.[12]

The major difference between Mercury and the Gusmobile was maneuverability. Gemini could fly, really fly. This was in stark contrast to its predecessor, so much so that Gus compared it to another love of his—cars. "Gemini's a Corvette. Mercury was a Volkswagen," he would often say. Gemini was "a whale of a lot more sophisticated" than Mercury. That's because Mercury was not very navigable. As Slayton admitted, the Mercury capsule was basically "designed for chimpanzees because they didn't trust man in it. We had to convert it a little bit to get a man backup [system]." Schirra, who flew in the Mercury program and would also fly in Gemini and Apollo, knew Mercury was not much of a flying machine. "I initially said they should take one of those circus performers they'd shoot out of a cannon because Mercury wasn't a 'pilot in space,' it was a 'human in space.' We didn't really contribute very much to the flight of the vehicle. We were lab specimens." But Gemini had to

be different. It had to be able to maneuver, to change orbits, to rendezvous and dock, in order for the moon to become a reality. And with Gus's
hard work, the Gusmobile was different, a true spacecraft. "He really
got the man into the loop. Mercury was sort of a program that proved
man could operate in space and survive," Slayton said. "Gemini was
basically a totally man-operated system. There was damned little on there
that didn't require a man in the loop. Gus was the guy who did most of
that." To help with maneuverability, the Gusmobile would also contain
an on-board computer, with fifteen kilobytes of memory, the first spacecraft to have one.[13]

In early 1962, in the midst of Mercury, NASA began testing the
Titan II missile, a very reliable Air Force delivery vehicle for nuclear
warheads that would be used as the booster. The Mercury Redstone
rocket used for the suborbital flight was weak, with just 78,000 pounds
of thrust; the Atlas had 360,000 pounds. The Gemini Titan II missile
had 430,000 pounds of thrust, capable of putting the two-man capsule,
which weighed four tons, into earth orbit.[14]

Everything was falling into place, and about a year after the conclusion of Mercury, NASA announced the first Gemini crews. As with the
later Apollo program, and for a variety of reasons, mission crews often
changed a number of times before the flights actually took place. The
plan was for Gemini to consist of ten to eleven flights, which would
require twenty to twenty-two astronauts to fill the seats. This is when it
became obvious that the New Nine were sorely needed, as well as additional groups in the near future, before Gemini got off the ground
because not every member of the Mercury Seven would continue in the
new program. Slayton himself was still medically ineligible. Carpenter
was still in the doghouse and would soon move on to deep sea exploration with the Navy. Glenn, though he wanted to fly in Gemini, was also
out because NASA, as well as the president of the United States, Jack
Kennedy himself, did not want to risk the life of a bona fide American
hero, so he retired in 1964. The first crewed Gemini mission would be
led by Alan Shepard, along with newcomer Tom Stafford, the first rookie

set to fly in space. Gus would be on the backup crew. The first mission, officially known as GT-3 (Gemini-Titan 3), would last eighteen full orbits, scheduled to launch in October 1964. GT-4 would last seven days, GT-5 would accomplish rendezvous and docking, and GT-6 would be a fourteen-day long-duration flight, which Slayton had initially pegged for Gus and Frank Borman. Other flights would follow over a two-year period. That was the preliminary plan anyway.[15]

As things were shaping up and crews were beginning their training, another original astronaut was lost, changing up the rotation. Shepard, set to kick off the Gemini program, just as he had the first Mercury, woke up one morning with a severe bout of dizziness, coupled with nausea and vomiting. Doctors determined he was suffering from Meniere's disease, an inner-ear problem. With such a condition, he was flying nowhere. He was grounded and off the flight rotation, and who knew for how long, as there was no medical procedure at that time to fix his condition. So, he joined Slayton in the astronaut office. Now Gemini 3 would fall to Gus. He was now at the head of the line.[16]

With the change in missions, Slayton changed up the crews, giving Gus a new pilot, rookie John Young. The crew for Gemini 3 was officially named in April 1964. Borman was moved to commander of Gemini 7, which would be the fourteen-day long-duration mission, which Slayton thought he could handle because he was "tenacious enough to stick it out." Gus and his new "space twin" jumped in with gusto to begin preparation to fly the first manned Gemini mission, which was trimmed from 18 orbits to just 3. The workload was downright brutal at times, tough on both the astronauts and their families. It was so much more rigorous than Mercury, Gus wrote, that it made it look "pretty soft" by comparison. In addition to their regular flight training, astronauts often flew across the country to visit the various facilities where both the spacecraft and the boosters were being built. They took 45 trips to McDonnell in St. Louis where they spent countless hours in meetings and in simulators. There was also simulator training at the Cape and, once the capsule arrived and was mated to the booster, the crew engaged in a number of tests in the cockpit itself. In addition, there were lots of hours in the classroom—"160 hours of geology, 50 hours of flight mechanics, and 20 hours of math review," Gus wrote. There was also time spent studying astronomy, rocket propulsion, computers, aerodynamics, communications, physics of the upper atmosphere, guidance and navigation, selenology, environmental control, and global meteorology. "If that sounds like the glamorous life," Gus said, "let me say it wasn't." The "days seemed to have forty-eight hours, the weeks fourteen days, and still there was never enough time. We saw our families just often enough to reassure our youngsters they still had fathers."[17]

As with any manned spaceflight program, before missions could take place, both the booster and the spacecraft had to be tested, just as they were in Mercury and would be in Apollo. The first two Gemini missions were unmanned tests of the Titan booster and Gemini spacecraft. Gordo Cooper had ended the Mercury program with his flight in May 1963. Gemini's first unmanned test, GT-1, was conducted nearly a year later, in April 1964. The test was flawless, sending the spacecraft into orbit, where it completed 63 revolutions of the earth. But before Gus could climb into his Gemini capsule, the second unmanned test, GT-2, had to get off the ground. It would be a simple suborbital flight to test the capsule's heat shield, a critical exam before it could be man-rated for astronauts. But as of December 1964, there had been one delay after another—lightning strikes and 2 hurricanes being the most frustrating instances. The original cost of Gemini was projected at $600 million; by December, NASA had spent $1.3 billion and hadn't flown a single manned mission.[18]

The second test was finally scheduled to take place on December 9. With the prime and backup crews present, along with members of the press, the engines ignited after the countdown reached zero but shut down three seconds before liftoff when the MDS—Malfunction Detection System—detected a problem and aborted the launch, just as it was designed to do. This meant another delay but Gus didn't brood as might be expected. He fully understood that in the business of flying in space, delays happen and sometimes they are for the best. It was the one business, he once said, where "failure means progress just about 99 percent of the time. So you learn patience." On January 19, 1965, GT-2 was tested successfully, paving the way for Gus to take Gemini 3 into space.[19]

And the hard work would pay off too, ensuring that Gemini 3 would be a complete success. "John and I had practically lived with our spacecraft since the first rivet was put into it at the McDonnell plant," Gus said. "We had studied every one of its systems as each was installed, and sweated out the glitches along with the McDonnell engineers. So we had the vital ingredient of confidence going for us all the way." Spending as much time as he had on the Gemini spacecraft, Gus knew the spacecraft "was going to work. It was as simple as that." It was now time to fly.[20]

But before the launch, Gus had a little bit of fun at the expense of his own misfortune with *Liberty Bell 7*. Since each spacecraft had been named in Mercury, naturally Gus thought it would be okay to name his Gemini craft. So, in a moment of humor, he decided to call it the *Molly Brown*, after the popular Broadway play *The Unsinkable Molly Brown* from Titanic lore, which was popular at the time. NASA brass did not find the name amusing in the least so they asked Gus if he had any other choices. And Gus being Gus, he replied, "How about the *Titanic*?" Realizing that Gus was probably not kidding around, the bosses yielded and *Molly Brown* it was, but after Gemini 3 the chiefs discontinued spacecraft names until Apollo 9.[21]

■   ■   ■

Gemini 3 launched on March 23, 1965, twenty-two months after the previous American spaceflight. Though it was a little later than NASA originally had hoped, everyone was hard at work making sure everything was in ship shape—even the secretaries. The day before the launch, Gus decided he didn't like the bed in the crew quarters at the Cape. It was too hard, and he needed a softer one, he said, but it was also too narrow, and he was afraid he might roll out during the night, hit the floor, and possibly hurt himself bad enough that he couldn't make the flight. Lola Morrow worked as a secretary in the astronaut office and had the respect of every man there, especially Gus. She was a go-getter. All the astronauts loved her and she them. She would handle the request. After Gus's complaint, though, John Young decided he didn't like his either. Nor did the backup crew of Wally Schirra and Tom Stafford. They all wanted new ones. So Morrow, who had been at NASA since 1962, called a furniture store, ordered four new beds, had Slayton sign the purchase order, and had them delivered that very day, assembled, and ready for the astronaut's bedtime at 8:00 p.m.[22]

After a better night's sleep, Slayton woke the crew at 4:40 a.m. They showered and had their final medical checks, which took only ten minutes. "They were feeling us to see if we were still warm and breathing, I guess," Gus said. Sam Piper, the steward at the Cape, prepared the crew a filling breakfast that included a two-pound porterhouse steak for each of them. On long missions, astronauts would learn to stick to a NASA-ordered low-residue diet in the days leading up to launch for obvious reasons, thus making the flight a little more enjoyable for the entire crew. Urinating in zero gravity is one thing; defecating is quite another. But since Gemini 3 would only last three orbits—a little over four hours—both astronauts could eat heartily.[23]

After breakfast, the crew suited up and were driven out to the launch pad. There the backup crew, Wally Schirra and Tom Stafford, were already in the spacecraft to set switches and get everything ready for the flight. In fact, Schirra and Stafford had been at the pad "most of the night," Gus said, perhaps the "two tiredest citizens at Cape Kennedy that morning." After one minor hold in the count, Gemini 3 roared off Pad 19 that morning at 9:24, and the crew heard the voice of Cape CapCom Gordo Cooper, "You're on your way, *Molly Brown*!" Project Gemini was off and running.[24]

In just under six minutes, Gemini 3 was in an elliptical orbit, with an apogee of 125 miles and a perigee of 87 miles. Upon reaching orbit, Gus became the first person in history to travel into space twice. It would be a simple "shakedown" cruise to test all the systems of the spacecraft to see if it was ready for the full battery of maneuvers required in the Gemini program. And with only three scheduled orbits, there was little time for much sight-seeing or fun and games. "I suppose I should have had some inspiring thoughts about being the first American to fly twice in space, but the truth was I was much too busy, as was John," he wrote in his book *Gemini*. This included talking to Betty. When NASA flight controllers allowed her to come to Mission Control and speak to her husband while he was in orbit, Gus refused to talk to her. He had too much to do. But Betty understood her husband and didn't really want to talk to him either.[25]

Gus did steal a few minutes to look out the window at the earth, something he had very little time for during his fifteen-minute suborbital flight during Mercury. The "scenery of space is magnificent," Gus wrote. It was hard to describe "the incredibly beautiful views we've seen during our flights." He even got a little spiritual with his assessment. Gus considered himself a Christian even though during the introductory press conference in 1959 he said he was "not real active in church." Nevertheless, he clearly believed in God and said so. It had been reported that Yuri Gagarin, the first man in space, had said that while in orbit he "looked and looked and looked but I didn't see God," though this has been disputed by Gagarin's family. Yet to Gus, God was everywhere. "If Major Gagarin found no evidence of God in space," he wrote, "he must never have looked out his cabin window."[26]

With only three revolutions of the earth, things were compressed into a pretty tight schedule. But soon after reaching orbit, "all hell broke loose." John Young, in the pilot seat on the right side of the capsule, was monitoring the oxygen gauges and noticed that the pressure had dropped sharply in both their suits and the spacecraft itself. A drop in the ship's oxygen pressure could kill them both. Gus instinctively pulled down the visor on his helmet. But, he later wrote, it suddenly struck him, "If the oxygen pressure is really gone, it won't make any difference. You've had it already." Luckily the difficulty was not a lack of oxygen but a problem with an electrical converter system that powered the spacecraft instruments. It had malfunctioned. Young switched on a backup system and all was well in the cockpit.[27]

Work soon resumed. On the first orbit, as they passed over the continental United States, Gus fired the spacecraft's thrusters for seventy-four seconds, which lowered the orbit, making *Molly Brown* the first spacecraft in history to change orbits, an essential requirement for rendezvous and docking maneuvers, which would come during a later mission. This was something that no Soviet spacecraft could do. On the second revolution, while passing over the Indian Ocean, Gus fired the aft thrusters and the nose thrusters, which changed *Molly Brown*'s orbital inclination by a fraction of a degree. On the last orbit, Gus fired the thrusters again to lower the perigee, which changed the shape of the orbit, making it rounder and less elliptical. The perigee, or closest point to the earth, was now just fifty-two statute miles, which meant that even if the retro rockets—used to slow down the spacecraft enough to re-enter the earth's atmosphere—failed to fire, *Molly Brown* could still re-enter safely, "possibly an orbit later."[28]

Overall, Gus was very pleased with the success of Gemini 3, calling it an "almost pure textbook flight," he wrote. "To our intense satisfaction we were able to carry out these maneuvers almost exactly as planned, confirming that our Gemini spacecraft was capable of rendezvous missions, in which changes of orbit and flight path are a requirement. To the best of our knowledge no astronaut or spacecraft had ever accomplished these maneuvers before." *Molly Brown* "flew like a queen." And the maneuvers were very easy to execute, owing to Gus's work on the spacecraft. "The procedure is simple," he said at a March 1965 press conference. "All we do is turn on the OAMS—the Orbital Attitude and Maneuvering System—and pull out the throttle I have on my left side, put the nose on the horizon, and start thrusting."[29]

Despite the heavy, compact workload during the three-orbit mission, there was one moment of amusement. John Young, the mission's newbie, carried out a very rookie-like maneuver. He brought contraband, in the form of a corned-beef sandwich, onto the spacecraft, tucked neatly away in one of the pockets of his spacesuit. Wally Schirra had picked it up from Wolfie's Deli in Cocoa Beach and given it to Young. Astronauts were strictly forbidden to do this but Young did have the permission of Slayton. Near the end of the flight, when the crew was supposed to be trying out their space food, which astronauts were famous for disliking, Young brought out the sandwich and asked Gus, "You care for a corned beef sandwich, skipper?" The comment surprised Gus. "If I could have fallen out of my couch, I would have," he said. Gus took a few bites. But when crumbs began to float around in the capsule, they decided to put it away. Crumbs and debris can easily get caught in switches and could threaten the mission, which is why there are strict rules about such items aboard any spacecraft.[30]

NASA officials were upset upon finding it out, as were members of Congress. Jim Webb had to promise the nation's representatives that the sandwich episode would never happen again. To guard against it, new, stricter rules were put in place. From that moment forward, everything being brought on-board had to be cleared by Slayton and nothing like a corned beef sandwich would ever be allowed again. Even though Schirra had picked up the sandwich and given it to Young, who brought it onto the spacecraft with Slayton's permission, Gus took the blame for it. "It was my fault that the sandwich got on board," he told Betty. "I should have known because I was in charge of the flight." For his part, John Young thought the "hubbub was completely unnecessary and blown totally out of proportion." He didn't think "it was any big deal," and admitted in his memoirs that it was actually the third sandwich brought on-board in the space program. There had been two in Mercury. Not to mention the fact that part of their mission "was to evaluate space food." He did receive a formal reprimand from Slayton but it did not affect his astronaut career. He went on to make five more spaceflights, including Apollo 16, when he walked on the moon, and the first Space Shuttle flight in 1981.[31]

After four hours and fifty-three minutes in space, Gemini 3 re-entered the earth's atmosphere. The spacecraft splashed down near

Grand Turk Island in the West Indies about forty-five miles away from the recovery vessels. *Molly Brown* did not sink but was certainly not a smooth ship. As Young later said, "It was no boat." Waiting for the choppers, sitting in the rocking capsule, Gus admitted his first thought was not a positive one. "Oh my God," he thought, "here we go again!" Peering out his window all he could see was water rather than blue sky, which is what he should have seen based on how Gemini was designed. He suddenly realized that he had not cut loose the parachute and the wind had caught it and was pulling *Molly Brown* across the sea. "Remembering that prematurely blown hatch on my *Liberty Bell 7*," Gus later wrote, "it took all the nerve I could muster to reach out and trigger the parachute-release mechanism. But with the parachute gone, we bobbed to the surface like a cork in the position we were supposed to take."[32]

After a half-hour of bobbing in the water, sitting in a hot and humid capsule, Gus must have thought the big breakfast and the corned beef sandwich a bad idea. The motion caused him to get seasick and throw up the sandwich and everything else he had in his stomach. John Young the Navy man was steady as a rock and managed to keep his food down. But despite being sick, Gus would not open the hatch until Navy divers had arrived and attached a flotation collar to the spacecraft. He wouldn't lose the *Molly Brown*. Gemini 3 was in the books.[33]

Soon after landing on the recovery ship, the USS *Intrepid*, President Lyndon Johnson phoned to congratulate them on the success of the mission, which, he said, "confirms again the vital role that man has to play in space exploration, and particularly in the peaceful use of the frontier of space." He even praised the spacecraft, noting that the "*Molly Brown* was as unsinkable as her namesake and we are all mighty happy about it." After speaking to both Gus and John Young, the president invited them to Washington on March 26, three days after the flight. At the White House that Friday, with Betty and the two boys present, Gus was awarded his second NASA Distinguished Service Medal. This one, though, was one of immense satisfaction because it was awarded to him personally by the president of the United States. Gus had also been awarded after his Mercury flight, but because of the mishap, he did not get a White House visit. This time, having his family with him and getting to spend some time with the president and first lady was as good as it could get. John Young was also awarded a medal, and both men and their families were given parades in Washington, D.C., New York, and Chicago. The public praise showed that the three-orbit flight was a crucial mission that would determine if Gemini could move forward. With Gemini 3 completed, NASA was one step closer to reaching the moon with Apollo.[34]

■   ■   ■

Now that the program was ready to go full speed ahead, it was getting a new home, at least for the manned aspect of it. Early on, all flight operations—six Mercury missions and Gus's Gemini 3 flight—were controlled and monitored from the Cape, which was first established as a testing ground for U.S. rockets in 1949. The area got its name—Cape Canaveral—from Spanish explorers who landed in the 1520s. After President Kennedy's assassination in 1963, the new president, Lyndon Johnson, changed the name to Cape Kennedy with an executive order, which lasted until 1973, when it was changed back to Canaveral after the outcry of native Floridians who lived in the area. But the space center at the Cape would become the John F. Kennedy Space Center (KSC).

The astronauts themselves had their headquarters—where their offices were located and where they trained—in Langley, Virginia. Gilruth's Space Task Group was in Beltsville, Maryland, at the Goddard Space Center, but it soon became obvious the facility wouldn't be able to handle anything beyond Mercury. NASA's first administrator, T. Keith Glennan, wanted the STG moved to the Bay Area. Gilruth, who had a passion for sailing, wanted it kept in close proximity to the Chesapeake Bay. With NASA moving into Gemini, then Apollo, both far larger and more complex programs than Mercury, Webb saw that an entirely new site would be needed to house everything in one spot. Chris Kraft wrote that a lunar landing program "would require engineering and test facilities far beyond anything at Langley. Astronaut training would be much more complex than our simple procedures trainer linked to mission control. And even the control center was inadequate. To manage and control missions to the moon, we'd need a new and bigger center, along with changes still unknown in the worldwide tracking network."[35]

On September 1, 1961, after a study conducted by a search committee headed by John Parsons, NASA announced the location of its new Manned Spacecraft Center—Houston, Texas. Now everything, except launch operations, which had to remain at the Cape, would shift a thousand miles to the west, and the new call-sign in astronaut communications with the ground would be the now all-familiar "Houston." This was not NASA's plan, however. The agency had recently purchased additional acreage near the Cape to construct a mission control center. At least publicly, the reason for the move was so that all operations would not be in the same locale, so communications could remain clear. Some, however, have contended that Houston was selected because of the massive power held by Texas lawmakers in the U.S. House of Representatives. After all, the powerful Speaker of the House, Sam Rayburn, was a Texan, as was Congressman Albert Thomas, who represented a district in Houston and chaired a House subcommittee that oversaw defense appropriations and also had a degree from nearby Rice University, and Congressman Olin Teague, who represented another Texas district and sat on the House committee that oversaw NASA, chairing the subcommittee on manned spaceflight. All three were certainly major players, but the real reason was then-Vice President Lyndon Baines Johnson.[36]

Before becoming Vice President, Johnson had been Senate Majority Leader and one of Washington's most powerful Senators throughout the 1950s, if not ever. LBJ biographer Robert Caro has called him the "Master of the Senate," and he truly was. But now he was in a new job that had very little real power to wield, making a man like Lyndon Johnson downright miserable. Kennedy, who picked LBJ as his running mate because he knew he needed to hold the "Solid South," including the vital state of Texas, to win the presidency, was sensitive to Johnson's ego and had even instructed one of his closest aides, Kenneth O'Donnell, to keep Johnson happy. One area where the vice president could do some good was space, since Johnson had such an avid interest in it. As vice president, Johnson was named chairman of the National Aeronautics and Space Council, an advisory position with no power to make any real decisions on space policy, just recommendations to the president, who "was not about to abdicate those decisions to anyone," as Jim Webb, who served with Johnson until 1968, would put it. Other members of the council included Secretary of State Dean Rusk, Secretary of Defense Robert McNamara, and Atomic Energy Commissioner Glenn T. Seaborg. Despite this plum, it seemed that Johnson was doing very little with what he had. The *New York Times* reported that at NASA, "Mr. Johnson's hand, if it has been laid upon that organization at all, has been light, indeed."[37]

But even if it may have seemed that way on the outside, reality was far different. Johnson was no simple, power-hungry, money-grubbing politico; he knew a lot about space and space policy. In fact, not only had he led the Senate investigation of the U.S. space program in the wake of Sputnik in 1957, he introduced the bill that created the very council that he now chaired as well as NASA itself. Johnson consistently pushed Eisenhower for a more aggressive space policy to challenge the Russians and beat them no matter the costs involved. And that's why Kennedy relied on him to help formulate his own space policy, as he did with the memo that outlined a race for the moon as the ultimate objective for America, giving Kennedy the ammunition to stroll confidently into the House chamber and proclaim to the world that the United States would land a man on the moon before 1970. It is an achievement for which LBJ gets little to no credit.

Despite his weakened office, Lyndon Johnson still knew how to get things done. His close friend and mentor, Speaker Rayburn, as well as Albert Thomas and Olin Teague, certainly had some influence and were a big part of the deal, but when it came to getting government largess for Texas, Johnson exercised more clout, much more. And he knew the businesses and corporations that could help get it done. "Any business leader who didn't know Lyndon Johnson," wrote one of LBJ's aides, "must have been on vacation."[38]

The site chosen for the new center was actually twenty-five miles south of the city of Houston, at Clear Lake, an area Norman Mailer described as a "flat, anonymous, and near to tree-impoverished plain which runs in one undistinguished and not very green stretch from Houston to Galveston." Two of LBJ's biggest supporters were heavily involved, with one of them getting the plushest of plums. Humble Oil, now ExxonMobile, a major Johnson donor and backer, agreed to donate a thousand acres to Rice University with the understanding that it would be given to NASA or, if it were not, that it would be returned to Humble. Rice gave the land to NASA and entered the budding space business by creating several degree programs in space studies and aeronautical engineering. For its part, Humble received a huge tax deduction, retained the oil and gas rights, and kept possession of significant land surrounding the new center, which they developed into suburbs. Another major Johnson supporter, Brown and Root, a massive construction firm headed by George R. Brown, a Rice alum who had helped Johnson craft his April 1961 memo to the president, received the contracts to build the facilities of the Manned Spacecraft Center, which is today known as the Johnson Space Center. The connections showed that LBJ's hand was all over the deal.[39]

An official narrative—*An Administrative History of NASA, 1958–1963*—written by NASA historians, noted, "There was considerable speculation that the selection of the Houston site was influenced by the fact that a Texan, Lyndon Johnson, was Vice President and chairman of the Space Council and that a Houston congressman, Albert Thomas, was the chairman of the House Independent Offices Appropriations Subcommittee, the subcommittee handling NASA appropriations."[40]

It was more than mere speculation. Florida Senator George Smathers, a close Kennedy friend, knew what had really happened behind the curtains. He was not happy one bit about losing a big chunk of the spaceflight apparatus from his state and blamed it all on LBJ. "He and I had a big argument about it, big fight. Johnson tried to act like he didn't know," Smathers said, referring to Johnson's habit of playing ignorant about controversial matters that involved underhanded dealings. "It never has made sense to have a big operation at Cape Canaveral and another big operation in Texas. But that's what we got, and we got that because Kennedy allowed Johnson to become the theoretical head of the space program."[41]

Gilruth could certainly grumble and gripe about having to leave cosmopolitan Virginia for rustic south Texas but there was little he could do about it. Initially, NASA's offices were spread around downtown Houston and at Rice University. It would be a few years before the new center was constructed at a cost of $193 million. Yet the eventual construction of the complex of buildings that would become the center would do little to brighten Gilruth's mood. Upon entering the gate and passing the guard shack, Mailer later wrote, "there was no way to determine whether one was approaching an industrial complex...or traveling into a marvelously up-to-date minimum-security prison." When looking over the site, Webb brought the political reality home to a less-than-impressed Gilruth. "Bob, what has [Virginia senior senator] Harry Byrd ever done for NASA?" The answer was a scant nothing, and Gilruth knew it. In fact, Byrd had been a persistent critic of NASA. But LBJ had done plenty. It would be hard to argue that any single person did more to advance the American space program and put a man on the moon than Bob Gilruth, but this was one fight he would never win no matter how hard he tried. LBJ was one old bull no one could tame. The new site would be in Texas, whether Bob Gilruth liked it or not.[42]

Texas and the rest of the South, due to a preponderance of congressional committee chairmen, might have gotten the lion's share of the space money flowing out of Washington, but every part of the country would get a piece of the action. There was simply no way to get the support of the requisite members of Congress without giving them some of the largess in return. And that largess was large indeed. At the height of its prestige, NASA accounted for nearly five percent of the entire federal budget. So, in addition to major facilities in Florida, Alabama, Mississippi, and Texas, thousands of contracts, some 20,000 during Apollo, would be spread far and wide—from California to Massachusetts. It was the price of doing business in the American Congress.

With Houston as the new center of the astronauts' world, they all relocated to the barrenness of south Texas. Yet with NASA in town, and a big checkbook in hand, civilization would quickly follow, with hotels, shopping centers, and restaurants aplenty. New subdivisions began popping up and the astronauts, flush with cash from their *Life* magazine deal, began building homes near the Manned Spacecraft Center. Most of the Mercury Seven had homes in the Timber Cove subdivision, across Taylor Lake from El Lago Estates, where most of the New Nine would buy lots and build homes.

Astronaut wives were an important part of the successful equation. And in some cases, they were very important indeed. With intense training and the accompanying travel schedule, astronauts were gone far more days out of the year than they were home, in some cases as many as 250 days out of 365. While training for his Gemini flight, Frank Borman spent so much time training that he "could find every switch, knob, and lever in the spacecraft blindfolded. But when I got home, I didn't know where Susan kept the water glasses." They had to rely on their wives for almost everything. When Jim Lovell, a member of the New Nine, was ready to build a home for his family in Timber Cove, he handed his wife Marilyn the blueprints for the house and headed to work. "You have to build the house," he told her. "I'm going to be too busy training." And she oversaw the building of a beautiful home, as did other wives who were given similar tasks. Author Lily Koppel has called them "unsung heroes," as well as America's first reality stars, and clearly they were. Gemini and Apollo astronaut Gene Cernan wrote that "all of those incredible wives should be in the history books."[43]

■ ■ ■

While their wives were acting as true homemakers, the astronauts saw their hard work continue unabated. For Gemini's second manned flight, Slayton chose Jim McDivitt and Ed White for the mission that would have a "big item" attached to it—a spacewalk. Initially, NASA, not wanting to rush things, had thought a limited spacewalk was possible on Gemini 5, a maneuver called a "stand-up EVA," where Pete Conrad, a New Nine astronaut who would be flying with Mercury veteran Gordo Cooper, would simply open the hatch and stand up but not leave the confines of the spacecraft while attached to a tether. Now that would have to change to a full EVA, and it would be moved up to Gemini 4. The reason was, as always, the Soviets. Five days before Gus lifted off with *Molly Brown*, a Russian cosmonaut named Alexei Leonov performed the world's first spacewalk, floating outside his Voskhod spacecraft for ten minutes. Grainy, black-and-white television footage of the event raced around the globe.[44]

But once again the Soviets had managed to use gimmickry to fool the world into believing they had a sound space program and were poised to win the moon race. To an unsuspecting world, the Soviets forged ahead; the reality was more like a Keystone Kops routine. For one thing, the Russian spacecraft was nothing new, and certainly not capable of a serious and safe EVA. It even launched without ejection seats in order to pull off the spacewalk stunt, thereby making for a much more dangerous mission. In addition, Soviet engineers had to create a makeshift, collapsible air lock, made of canvas and attached to the hatch, for Leonov to get out and into the void. The reason for this was the primitive, out-of-date systems in the Soviet spacecraft. The Russians used older glass tubes in their electrical system, which were prone to overheat, so exposing them to the cold of space might cause them to explode, crippling the spacecraft. As a result, Soviet cosmonauts could not open their hatch. Hence the air lock. It was attached to the fuselage and would be inflated with air, then the spacecraft could be depressurized, and the hatch would be opened into the air lock.

Leonov would enter the air lock and close the spacecraft hatch. He could then open the outer hatch of the air lock and float into space.[45]

For about ten minutes Leonov floated freely, attached only to a tether, over the Soviet Union in order to be in close contact with ground controllers. After proving their point, he was told to get back in. But because of the crude nature of the air lock, he nearly died trying. His suit was overinflated, and he simply could not fit. So, he had to do something very dangerous: use a valve on his spacesuit to release some of the oxygen, deflating the suit and allowing him to maneuver himself back into the air lock and into the spacecraft. Once the hatch was closed and the spacecraft re-pressurized, the air lock was jettisoned.[46]

Upon re-entry, things were no better. Whereas the American space program made use of the U.S. Navy to pluck astronauts from the sea, where the water made for a much softer landing, Soviet cosmonauts, because of the vast territory of the USSR, came down on land, a much rougher landing. The Voskhod 2 spacecraft, though, crashed into trees in the Ural Mountains, wedging itself so tightly that the crew could not open the hatch to get out. They had to spend a frigid night waiting for rescue crews to get to them. A helicopter did fly overhead, located the spacecraft, but could not land because of the thick forest. The next day, both were rescued safely. But once again, these stories did not make the nightly news broadcast in Moscow or anywhere else in the world. The only thing anyone knew was that the Soviet Union had achieved yet another first in space—EVA—as risky and dangerous as it had been.[47]

With Leonov's spacewalk, NASA knew they had to answer and do it quickly. Even with the flight scheduled for June, it "wasn't until May 25 that everybody signed off on the EVA with the tether," Slayton wrote. EVA, or spacewalking, was one of the necessary steps on the road to the moon, one on a list of tasks that Gemini had to conquer. Astronauts had to be able to work outside the spacecraft, either while in motion or on the surface of the moon. Landing on the lunar surface to simply look out the windows wasn't very appealing. A spacewalk was every bit as much a test flight as piloting the latest spacecraft. The primary equipment being tested, though, was the spacesuit. Space is the most inhospitable environment in the known universe. There is no air, but there are high levels of radiation and violent extremes between hot and cold. With no atmosphere, the moon also has wild temperature variations of 400–500 degrees Fahrenheit. Standing in directly sunlight, the temperatures can rise to 250 degrees, but step into the shadow of the lunar lander, just a few feet away, and it plunges to 200 below or more. It would be the same on a spacewalk in earth orbit, where direct sunlight would push temperatures to 250 degrees, while the shadow of the sun on the other side of the spacecraft would plunge it to minus 150 degrees. So the spacesuit would be lined with ten extra layers of protective material to insulate an astronaut from both extremes at nearly the same time. If the suit failed—most likely from a puncture by a micro-meteorite—an astronaut's blood would boil or freeze within seconds. The cold of space was so severe that steel could freeze to the point where it was as brittle as glass. In spite of the grave risks involved, Ed White would be the American program's guinea pig for the first step into the void.[48]

■   ■   ■

Everyone who knew Ed White liked and respected him. You might even say they loved him. And it seemed that no one ever had a bad thing to say about him. "I don't know of any astronaut who was more genu-inely liked and admired," wrote Frank Borman. This was true through-out his flying career, even in his earliest days when pilots tended to get a little rambunctious. James Salter, a West Point graduate and Air Force flyer who left the military to become a writer, knew Ed during those youthful days as a fighter pilot. He had been "the first person I met when I came to the squadron and I knew him well," Salter later wrote. "He had a fair, almost milky, complexion and reddish hair. An athlete, a hurdler; you see his face on many campuses, idealistic, aglow. He was an excellent pilot, acknowledged as such by those implacable judges, the ground crews. They did not revere him as they did the ruffians who might drink with them, discuss the merits of the squadron commander or sexual exploits, but they respected him and his proper, almost studious ways. God and country—these were the things he had been bred for." Ed was "a man who could be relied upon—in every way," Salter said, and those around him knew they were "intimate with greatness." Ed White would achieve great things, Salter believed.[49]

Ed's two children, Ed White III and Bonnie Lynn, remembered their father as "the very picture of an American hero. He was good-looking, athletic, and tall—at least for an astronaut. He was a thoughtful man of faith, and many thought one day he'd go into politics. He certainly seemed like a natural on camera."[50]

Wally Schirra, who would command the Apollo 1 backup crew, also thought highly of Ed, looking at him as "a real candidate for superstar!" who would probably retire as a general with three or four stars on his shoulders. Another member of the backup crew, Walt Cunningham, wrote that "Ed is a golden boy." He was "meticulous, tall, clean-cut, and a fierce advocate of all the basic virtues: God, country, Mother, and religion." John Young said there was "no question" about it, Ed was "as capable as they came." Michael Collins, who graduated from West Point in the same class as Ed, called him a good friend who was "easy to work with." But, he added, "Ed also had a deep, serious side: he was a proud practicing patriot, and was not ashamed to say so." To Buzz Aldrin, who called Ed "my close friend," he was a "sharp pilot and a thoughtful man who made friends easily." Aside from Aldrin, Ed was close friends with several in the astronaut program, including Jim McDivitt, who said Ed was "the best friend I ever had," and Neil Armstrong. They became very close friends. "Ed and I bought some property together and split it. I built my house on one-half of it, and he built his house on the other. We were good friends, neighbors," Armstrong said. Even the neighborhood kids thought a lot of Ed White. "Of all the astronauts we knew," said one, "Mr. White was the nicest."[51]

Perhaps out of everyone in the astronaut corps, though, Ed's closest friend was Frank Borman, who had been two classes ahead of Ed at the Point. "With one exception, I regarded the other astronauts more as

professional comrades than as truly intimate friends. The exception was
Ed White." He "might as well have been the brother I never had," Borman wrote in his autobiography. Ed "really was the astronaut's astronaut, a handsome and powerfully built man who actually seemed
indestructible." But he was also "a man of gentle strength and quiet
humor." They lived close to one another and their "shared philosophy
brought us close together," Borman said. "He and I were very much alike
in our devotion to our families." A deeply serious man, Borman didn't
socialize with other astronauts, not even his good friend Jim Lovell, with
whom he flew two missions, but with Ed things were different. He knew
him as well as anyone.[52]

But it was not because of Ed's immense likability that he was
tapped for such an important mission as Gemini 4. One reason, aside
from his skills as a pilot and engineer, was because he was such a phenomenal athlete and in superb shape, which might be useful for the
first EVA. And Borman knew this aspect of Ed very well. His "devotion
to physical conditioning drove those with more sedentary habits absolutely bonkers." Ed would jog three miles every day "without even
breathing hard, then would play two or three hours of handball or
squash. This was anathema to guys like Schirra and Armstrong. Neil
once commented, after watching White go through one of his conditioning regimes, that man had been allotted a finite number of heart-
beats in a lifetime and there was no need to speed up the process." But
for Ed it was a way of life and it was another feather in his cap that
earned him the respect of NASA executives and eventually put him on
the second manned Gemini flight.[53]

Early on Slayton had been thinking about putting Ed on whichever
mission conducted the first EVA. To fellow Gemini astronaut Gene
Cernan, Ed was "slender and good-looking and straight as an arrow,"
"an All-American, clean-cut...poster boy for the program. He was our
Yuri Gagarin. Ed was damned good and we could have picked no better
person to be the first American to walk in space." And "clean-cut" was
an understatement. Whereas many in the astronaut corps were prone to

chase women who were not their wives and race fast cars around the Cape to blow off steam and soothe their egos, Ed White never partook in any of those activities. He was a serious astronaut and the best husband and father one could wish for.[54]

Edward Higgins White II was born to fly. Other astronauts might have stumbled upon the occupation or were exposed to it later in life, but not Ed. He was, by his own words, "actually born into flying." And he was good at it. His closest friend during his Air Force days, Buzz Aldrin, who was a year ahead of him at West Point, said he was "a hell of a pilot." Having the most influence on his interest in flying was his father, Edward Higgins White Sr., a 1924 graduate of West Point who served in the Army Air Corps before it became a separate force apart from the Army in 1947. The Air Force Academy was started in 1954, and Ed's younger brother, James B. White, graduated in 1964, but was tragically killed in 1969 during the Vietnam conflict when his plane crashed in Laos. But before the creation of the Air Force Academy, prospective Air Force pilots went to West Point. Ed Sr. served in the Air Force for 33 years, flew over 100 different types of aircraft, conducted the first aircraft-to-train transfer of U.S. mail, and amassed over 8,000 hours of flight time before his retirement in 1957. He won two Legion of Merit awards and a Distinguished Service Medal.[55]

Service in America's military forces was not limited to Ed's father; the tradition ran strong in the White family. One of Ed's uncles, James White, attended West Point and served in the infantry, while another uncle, John White, graduated from the Naval Academy, served in the Marine Corps, and, as an embassy guard in Beijing in 1941, was captured by the Japanese and spent the entirety of World War II in a prisoner-of-war camp.[56]

Born in San Antonio, Texas, on November 14, 1930, Ed White II was destined to follow in the footsteps of his family and serve his country. More specifically, he would follow his father into the skies. At twelve years of age, his father took him for his first airplane ride in a T-6 trainer. He was, he said, "barely old enough to strap on a parachute," yet his father "allowed him to take over the controls of the plane." Other kids that age might have been intimidated and even scared by such a move, but not Ed. To him, "it felt like the most natural thing in the world to do."[57]

Being a military brat, Ed moved around the country as his father was stationed in different locales. He spent his final three years of high school at Western High in Georgetown, outside Washington, D.C., where he graduated in 1948. Iris Coopersmith, a former high school classmate of Ed, remembered him as "a very good looking young man with red hair and brown eyes. He was well liked and very good at his studies." Another classmate, Hank West, who also attended West Point with Ed, recollected that he had "a thirst for adventure, an unrelenting persistence in all pursuits, and a fierce competitiveness."[58]

Like his father and uncles before him, Ed knew that he would attend West Point, noting that "there never seemed to be any question that I would go there too," he wrote in *Life* magazine in 1965. He was nominated for appointment by Oklahoma congressman Ross Rizley and reported to the campus on July 15, 1948, for plebe summer, a grueling training program required of all incoming freshmen to acclimate them to life in the service academy.[59]

And there at West Point he showed himself to be the phenomenal athlete he was known as. He played soccer and ran track. Buzz Aldrin, who was at the Point for much of the same period of time, graduating one year ahead of him, often sat with Ed at the track table during meals. For a "lanky guy without an ounce of extra fat," Aldrin wrote, "Ed White could eat a lot of food," owing to his sky-high metabolism, something common in top-notch athletes. Walt Cunningham once joked that Ed could easily eat his own weight in seafood. Gus would also observe Ed's prodigious appetite, once noting that Ed "regards a porterhouse steak merely as openers." While on the West Point track team, Ed ran the 400-meter hurdles and set a school record. He was, in fact, so good that he nearly made the 1952 United States Olympic Team, missing the cut by less than a second. Physical training and proper diet would remain a staple for Ed White throughout his life.[60]

Ed also met his future wife, Patricia Eileen Finegan, whom Lily Koppel described as a "delicate-as-porcelain blonde," during a football weekend at West Point. Ed graduated in 1952, 128th out of 523 cadets, with a bachelor of science degree, but he ranked first in his class in physical fitness, not a surprise to all who knew him. The following year while at Bartow Air Force Base, his first posting, he married Pat. They eventually had two children—Ed White III, who was born later in 1953 and Bonnie Lynn, born in 1956. Soon after leaving West Point, Ed entered training in the U.S. Air Force and earned his wings, after which he was sent to Germany, where he flew the F-86 Sabre and the F-100, the newest Air Force fighter jet. Germany was where he got the idea that he wanted his flying career to move into space.[61]

Returning to the United States after three-and-a-half years in Germany, Ed's next step was to get his master's degree in aeronautical engineering at the University of Michigan, where he started in the fall of 1957. There he met someone who would become a major part of his astronaut career, Jim McDivitt, another Air Force pilot with dreams of spaceflight. He completed his degree in two years and, in 1959, began the next phase, the Air Force Test Pilot School at Edwards Air Force Base in California, where he met up with McDivitt once again. After finishing test pilot school, Ed was stationed at Wright-Patterson Air Force Base in Ohio, where Gus had also worked, and began test flying different types of aircraft. While at Wright-Patterson, another duty of his was to fly the planes used to train NASA's Mercury astronauts in weightlessness. Nicknamed the "vomit comet" because of its tendency to cause nausea and its typical side-effect in most trainees, the cargo-type planes, usually C-131s or C-135s, would fly in a rollercoaster, arcing motion—"flying parabolas," John Young called it—plunging straight down from its apex to simulate zero gravity for 20–25 seconds. A typical training flight would take the plunge 30–40 times. "I flew the big Air Force cargo planes through weightless maneuvers to test what would happen to a pilot in zero gravity," Ed said. "Two of my passengers were John Glenn and Deke Slayton who were practicing weightless flying for Project Mercury. Two other passengers of mine were Ham and Enos, the first Americans [chimpanzees] to fly in space. I'm able to kid that I've already gone weightless 1,200 times. This adds up to five hours of weightlessness, enough for three orbits of the Earth." Such unique flying experience would only add to what was already a stellar resume when NASA called for a new group of astronauts in 1962.[62]

Once selected, the "New Nine" were sent to Houston for their own introductory press conference, set for September 17, 1962. They flew in anonymously and were ordered upon arrival to check into the Rice Hotel under the name "Max Peck," which was the name of the hotel's manager. NASA thrived on secrecy, as it always did in those days. Before long, all nine "Max Pecks," bored of sitting in their rooms, were in the hotel bar where it didn't take them long to find each other and begin hobnobbing. John Young introduced Ed to Jim Lovell, who saw "something vaguely familiar" in his fellow astronaut's face. He had met him somewhere before, he was sure of it.

"Good to meet you," Lovell said, as the two shook hands.

"Actually, we've met already," Ed replied. But he was talking about a phone call he had placed to Lovell's room earlier that day. "I'm the Max Peck who called your room," Ed said. But Lovell just knew they had met in person before.

After the talk turned to their time at their respective service academies, Ed at West Point and Lovell at Annapolis, Lovell mentioned that he had attended an Army-Navy football game during his years at the Naval Academy and had swapped cuff links with an Army cadet. Both men were astonished to learn that it was Ed White who had given Jim Lovell his cufflinks a decade before. Lovell still had them.[63]

Ed's space career was soon off and running, and in October 1964 he was named to the Gemini 4 crew, a mission that would likely fly the following summer of 1965. And since Slayton had paired Ed with his old pal Jim McDivitt, who would be the mission's commander, it was a solid match. From their days together at Michigan and test pilot school, their friendship had grown a "bond between us like a band of steel," Ed wrote after their mission.[64]

As with Gus on Gemini 3, the Gemini 4 crew wanted to name their spacecraft, and they chose *American Eagle*, but NASA wouldn't allow it. Although the spacecraft would not have an official name, nor the flight have an official logo, the crew decided to put the American flag on their spacesuits, a first for NASA, and the tradition remained and was carried forward on all future space missions. In addition to Ed's EVA, it would also be the longest American spaceflight to date. Up to that point, the lengthiest U.S. mission was Gordo Cooper's Mercury flight, which came in at a little more than thirty-six hours. Gemini 4 would stretch four days. The big-ticket item, though, was Ed's spacewalk, which, in NASA tradition, was kept secret until almost the last minute. The spacesuit was not certified ready for flight until ten days before the launch, and the EVA was not publicly announced until a week before liftoff.[65]

■   ■   ■

Gemini 4 lifted off Pad 19 on June 3, 1965, at 10:16 a.m. The mission, with its historic firsts of a spacewalk and a four-day duration, was significant in other ways as well. The launch was broadcast live to Europe, as well as to Americans at home, via the Early Bird Satellite, another innovation that the Soviets did not have, showcasing yet again that they were nowhere near America in the overall space race. Because of the mission's long duration, this was the first use of a three-shift rotation of flight controllers to monitor every aspect of the mission, the brainchild of Chris Kraft, who designed the concept of Mission Control itself. Now that missions would be stretching into days, and even as long as two weeks, there was no way a single team of controllers could handle it alone, as it could during Mercury and for Gemini 3. So each mission control team would each serve an 8-hour shift led by a flight director.

Within 8 minutes of liftoff, the spacecraft entered an elliptical orbit of 100 miles by 175 miles. McDivitt quickly turned the spacecraft around to find the second stage booster, then several hundred feet away after being jettisoned. Even though this mission was not set to conduct a rendezvous procedure, McDivitt tried to use the spent stage as a target vehicle but it kept moving further away from the spacecraft no matter how much he maneuvered. After four attempts, and the use of a lot of maneuvering fuel, the crew threw in the towel. NASA still had a lot to learn about orbital mechanics at this stage of the program. Things moved and reacted differently in space than they did in the friendly skies on earth. Using the spacecraft's thrusters to accelerate toward an object increases the velocity, which places the vehicle in a higher orbit, whereby you will actually be "going slower than you were when you fired your thrusters to increase your speed," Slayton observed. "It's a hard thing to learn, since it's kind of backward from anything you know as a pilot." As Gemini and Apollo astronaut Gene Cernan later wrote, "The laws of orbital mechanics are about as strange as the regulations of the Internal Revenue Service." But there would be other opportunities to perfect those maneuvers in later missions. Gemini 4 had bigger and more dangerous fish to fry.[66]

The crew soon began preparations for Ed's spacewalk. Originally planned for the second orbit, McDivitt asked Mission Control that it be delayed until the third go-around in order for the crew to have a little extra time to prepare "because it was our first step in space and we wanted to be sure that the procedures were done thoroughly and correctly." NASA wanted the EVA to be on the first day early in the flight when Ed was still fresh and before he became fatigued during the four-day mission. This way if anything went wrong, he would have plenty of strength to handle the situation, so the third orbit was readily okayed.[67]

On the third orbit, at 2:34 p.m., about four and a half hours into the mission, both crew members were locked tightly in their spacesuits, breathing 100 percent pure oxygen, when the cabin was depressurized and Ed opened the hatch and stood up. He attached a color movie camera to the outside of the spacecraft so that his historic walk in space could be fully documented. Connected to a tether that supplied the suit with oxygen and provided a communications link, Ed gently floated out of the spacecraft and into the void of space, armed with a hand-held maneuvering unit (HHMU), called a "zip gun" that was fueled with pressurized oxygen to propel him through space. He also had a camera attached to the front of his spacesuit so he could take photographs while on his walk.[68]

Once outside, Ed floated as far away from the spacecraft as the tether would allow, using his gun to move when and where he wanted to. As he tumbled around in space, the camera caught a thermal glove gently floating out of the spacecraft. When the gun ran out of gas, Ed maneuvered under his own power, controlling his own movements, pulling himself around by the tether. With no air in space, there was absolutely no resistance on the spacecraft, and this was fully demonstrated during the spacewalk, so much so that Commander McDivitt said that whenever Ed pulled on the tether, he could feel the spacecraft move. Like those watching the television broadcast, McDivitt marveled at Ed's spacewalk. "It looks beautiful!" he said. "I feel like a million dollars!" Ed replied. He was having the time of his life, checking out the "spectacular" and "indescribable" views and snapping pictures. The experience made him feel very patriotic. "I felt red, white, and blue all over," he told *Life*. "At that moment I was a very lucky American."[69]

The entire EVA lasted all of twenty minutes, but that was twice as long as Leonov's excursion. The only problem Gemini 4 encountered was a communications issue during the spacewalk. This was a time before the advent of the worldwide satellite system whereby communications could be relayed around the globe to a central command center, so that if a spacecraft was in orbit on the other side of the planet, Mission Control in Houston could continue to talk to them. In the early days of the manned space program, NASA had to maintain tracking stations at several points around the world. Each had its own Capcom. Gus would be the Mission Control CapCom, the main one for the entire Gemini 4 mission, which relayed information between the crew and Chris Kraft, the flight director. Yet there were gaps in the communications network, short periods of time where there was no direct link to the spacecraft.

Once Gemini 4 moved out of range of the Hawaii tracking station and moved into the scope of Houston, Gus began calling the spacecraft. But he received no answer. He tried several times. Still no answer. After a few moments, Ed called to Mission Control. "Gus, I don't know if you can read us but it looks like we are over Houston," he said. But Gus couldn't hear anything. Members of the media tried to make an issue of this later on, that Ed had somehow been caught up in the moment and lost track of time, lost track of his mission parameters, lost track of his commander, and, most importantly, lost track of Mission Control. To this Slayton had only one word: "horseshit." None of it was true. Ed wasn't lost in the "euphoria" of his experience. It was simply "a communication problem: Gus could talk to Jim in the Gemini, but not directly to Ed, who was connected through a line in his tether. Ed couldn't hear Gus, and even Jim wasn't hearing everything. There was no other problem." Furthermore, once Ed did hear the message that was finally relayed to him by Chris Kraft himself, "The flight director says get back in," he immediately complied, although he said later that he felt "a certain sadness" that it was coming to an end. He did have a little difficulty maneuvering back inside to his seat in the spacecraft and a brief moment of trouble getting the hatch closed. But everything eventually came to a very successful conclusion.[70]

"It just felt plain normal," Ed wrote in an article for *Life* magazine after the flight, in echoes of his first trip in an airplane with his father years before. "From the moment I propelled myself away from Gemini 4, my strongest feeling was that of doing something that I had been trained for. There was absolutely no sensation of falling. The sensation of speed was the same as it was inside the capsule—and that was similar to the sensation you'd get flying over the earth in an aircraft at about 20,000 feet. There was no feeling at all of being in a hostile environment," he said.[71]

The mission ended with a successful splashdown four days later, completing NASA's longest mission to date. Ed White and his amazing feat garnered international attention, making him a star. "Not surprisingly, it was Ed's spectacular accomplishment that got most of the headlines," Gus wrote. And it was spectacular indeed. Leonov had "simply tumbled at the end of his tether," Gus noted, but Ed "actually controlled his own movements and so proved that man could function meaningfully outside of the spacecraft environment." And Ed was excited about what he had accomplished, so much so that once on board the recovery ship, the USS *Wasp*, he actually danced "a jig on the flight deck," wrote his friend and fellow astronaut Frank Borman, to the surprise of many. "The next day, White saw a few Marines and midshipmen having a tug of war and joined them. Quite a guy was Edward Higgins White II."[72]

Much of the country and the world thought so too. Vice President Hubert H. Humphrey, now the head of the Space Council, noted that the "public acclaim was impressive," especially for Ed White and his spacewalk. The more the general public saw of Ed, the more likeable they found him. In the post-flight press conference, Ed opened up about his religious views, revealing that he brought a few spiritual icons on board— a gold cross, a Star of David, and a St. Christopher medal. "I had great faith in myself and especially in Jim and I think I had a great faith in my God," he said. "So the reason I took these symbols was that I think that this was the most important thing that I had going for me." A devout Methodist, religion was as important a part of Ed's life as physical fitness and explains why he was such a kind husband, a loving father, and a loyal friend.[73]

Within days of the mission, President Johnson invited the Gemini 4 crew to the White House to be honored for their historic accomplishment, calling them the "Christopher Columbuses of the Twentieth Century." The president awarded them NASA's Distinguished Service Medal and promoted both to Lieutenant Colonel. While in Washington, they addressed a joint session of Congress and attended an event at the State Department where they showed the film of the mission, particularly the spacewalk, while narrating it for the assembled guests in the auditorium. The McDivitts and the Whites stayed the night at the White House, ate dinner with the president and first lady, and all enjoyed a swim in the White House indoor pool.[74]

If that were not enough, for better or worse there was plenty more to come. Vice President Humphrey wanted the Gemini 4 crew for another important endeavor—a public relations mission to counter the Soviet Union in the propaganda war in the Space Race. The Paris Air Show was underway, and the Soviets were putting on quite a display, showcasing their biggest hero, Yuri Gagarin, and also Alexei Leonov, for all the world to see. Yet the exhibition sponsored by the United States was less than impressive. Humphrey got the idea to take America's newest space heroes, who he said were "special favorites of [his]," to Paris to at least try to even the score. But he ran into opposition from NASA and specifically Jim Webb, who "opposed astronauts' travel for that purpose." But Humphrey thought NASA's "official" position was wrong, so he went to the president about it. Johnson overruled Webb and before long Humphrey, along with Ed White and Jim McDivitt and their wives, were on their way to Paris as goodwill ambassadors where they were a big hit, completely overshadowing the Soviets. Ed even met and shook hands with Leonov.[75]

Upon their return to the United States, tributes continued to pour in. They received a ticker-tape parade in Chicago and were both awarded honorary doctorates in aeronautical engineering by their alma mater, the University of Michigan. "I can hardly get used to people calling me colonel," Ed said. "I know in a million years, I'll never get used to people calling me doctor." Both men were awarded the Arnold Air Society's John F. Kennedy Award. Ed won the Society of Experimental Test Pilots Award for his spacewalk and the American Academy of Achievement's Golden Plate Award. He also paid a visit to West Point in the fall and spoke at halftime of a soccer match to rousing applause.[76]

■   ■   ■

Accolades and perks for their newest heroes aside, NASA was soon back in business flying, and there were still plenty of tasks remaining in

Gemini in order to move on to Apollo. Both Gus and Ed would be called upon to command backup crews for two of the most important missions—Gus would be the backup commander for Gemini 6, the rendezvous flight, while Ed would be the backup commander for Gemini 7, slated for fourteen days, NASA's longest-duration mission until Skylab in 1973.

The next mission, though, was Gemini 5, which would tackle long-duration and attempt to double Gemini 4. NASA believed that the mission to the moon might stretch as long as two weeks. Not wanting to push that hard just yet, Gordo Cooper and rookie Pete Conrad would extend Gemini to eight days, utilizing a new technology for the first time—the fuel cell. Battery operated spacecraft could only last a short time in earth orbit but fuel cells generated their own electrical power by combining liquid hydrogen and oxygen, producing power for the spacecraft and water for both consumption and cooling the electrical systems, all of which made it possible to remain in space long enough for a trip to the moon. Cooper and Conrad also came up with the idea for an official mission patch, to be worn on their space suits. Hailing from Oklahoma, Commander Cooper decided to use a Conestoga wagon with the slogan "Eight Days or Bust." NASA nixed the slogan though, fearing that if the mission did not make it eight full days it would be viewed as a failure, so it was not included on the official mission patch. The mission lasted seven days, twenty-two hours, and fifty-five minutes, just short of the mark. In the cramped space, Pete Conrad called it "eight days in a garbage can."[77]

The next objective was perhaps the most important—rendezvous and docking. Gemini 6, with Mercury veteran Wally Schirra as mission commander, backed up by Gus, and rookie Tom Stafford as pilot, would fly into orbit in October and rendezvous with an unmanned Agena target vehicle, which would be boosted into orbit atop an Atlas rocket. Gemini 6 would follow soon after, then find the target, fly in formation with it, and dock. With the crew strapped in their spacecraft, the Agena lifted off the pad but was soon lost on radar. Flight controllers determined that it had exploded, as a tracking ship began to track pieces of debris falling into the Atlantic Ocean. With no target with which to rendezvous and dock, Gemini 6 was momentarily scrubbed.

NASA soon devised a unique solution to the problem. Rather than cancel Gemini 6 altogether, or delay it until another Agena could be ready to go, they decided to push the schedule back to December, launch Gemini 7 first, with the crew of Commander Frank Borman, Ed as his backup, and pilot Jim Lovell, two New Nine astronauts. Gemini 6 would follow eight days later, renamed Gemini 6A because of the change in mission rotation. With two spacecraft in orbit at the same time, a rare feat, Gemini 7 would be used as a target for Gemini 6 but without the docking aspect of the mission, as there was no apparatus for the two spacecraft to merge. Docking would be bumped to a later mission. Gemini 7's official mission was long-duration, stretching to 14 days. That would still be the objective. It now had an additional function. With hard work and swift engineering by NASA technicians, Gemini 6 was removed from Pad 19, and Gemini 7 was put in its place. It lifted off on December 4, 1965. Once Gemini 7 was in orbit, Gemini 6 would be returned to the pad. The engineering team pulled off what could only be described as a technological miracle.

With Borman and Lovell in space, Gemini 6 was set to fly on December 12. Just before 10:00 a.m., there was an ignition but within one-and-a-half seconds the engines shut down. The clock had started but there was no liftoff. Events like these generally send shivers down everyone's spine, crew and Mission Control alike. But Commander Schirra, a veteran pilot relying on his instincts, did not eject the crew from the spacecraft. Unlike Mercury, and later Apollo, where the capsule had an escape tower that would ignite and pull the spacecraft away from the booster on the first sign of trouble, Gemini had ejection seats, like in a jet fighter. The commander could pull a D-ring beneath his seat to activate the ejection system. Doing this would likely have injured the crew, perhaps severely, as most ejections did, and would certainly have destroyed the spacecraft, wrecking their mission entirely. By playing it cool, Schirra saved Gemini 6. The problem was an electrical plug that had come loose and a dust cover that had been mistakenly left in place within the engine, blocking the flow of fuel. The MDS recognized the problems and shut down the engines, just as it was designed to do.

Three days later, on December 15, at 8:37 a.m., Gemini 6 finally lifted off, heading for orbit and a rendezvous with Gemini 7. By putting Gemini 6 in a lower orbit than Gemini 7, they were moving faster and could catch up. This is where the Gusmobile's full capabilities really started to pay off. After several maneuvers and thruster burns, Gemini 6 finally caught up to Gemini 7. During their rendezvous, both spacecraft maneuvered to within one foot of each other. Though there had been other areas of American superiority in space, NASA really began to pass the Soviets with rendezvous. Though the Russians claimed to have already conducted a rendezvous mission, it was yet another Soviet lie. They had put two spacecraft in orbit at the same time, which was a world first, but there was no rendezvous. The two craft found each other on radar, and maneuvered to within several miles of each other, but that was all they could manage. *Pravda* announced it as the world's first rendezvous in space but NASA knew better.

Controversy aside, Gemini 6 returned to earth the next day, while Gemini 7 remained in orbit for a few more days to set the space endurance record, yet the mission fell just short of fourteen days, about five and a half hours short to be exact. But after 206 orbits of the earth, and serving as the target for Gemini 6, Gemini 7 had done enough for one mission. Although doctors questioned whether the crew would be able to withstand such a long-duration flight in zero gravity, both Borman and Lovell were fine after what spending what amounted to two weeks in the front seat of a sports car. Lovell later called it "fourteen days in a men's room."[78]

Next up to bat was Gemini 8, also with a crucial mission parameter—docking. Commanding the mission would be Neil Armstrong, along with pilot Dave Scott. The flight was scheduled to last three days, dock with an unmanned Agena, and included a lengthy two-hour EVA by

Scott. Yet mission lasted less than eleven hours, included no spacewalk, and nearly ended in disaster. Armstrong successfully docked Gemini with the Agena, a first in space history, but soon after, the spacecraft began to spin. Thinking the problem was with the Agena, Armstrong undocked, but that only made things worse. A stuck thruster was causing the craft to tumble end over end. The crew was close to passing out but Armstrong fired the re-entry rockets, which stopped the spin, but also caused them to end the mission prematurely. They landed way off course in the Indian Ocean. Yet because of Armstrong's cool head, he saved the crew and the spacecraft. NASA knew he would be an astronaut they could rely on in the future.

The remaining missions would rendezvous and dock with an Agena, then perfect the fine art of EVA. Ed had conducted a flawless spacewalk, but he had not performed any tasks while outside the spacecraft. The remaining missions would prove that working in space was also possible. And this is where things almost ground to a halt.

Before the flight of Gemini 9, the original crew of Elliot See and Charlie Bassett, along with backups Tom Stafford and Gene Cernan, flew to St. Louis to check out the spacecraft. On February 28, 1966, coming in to land in dense fog, the T-38 piloted by See, with Bassett in the rear seat, slammed into a hanger building, instantly killing both men. This radically changed up the rotation, promoting the backup crew to the prime slot. The new mission would be renamed Gemini 9A. Stafford would be the first member of the New Nine to fly twice. In fact, four members of the group would fly a second time in Gemini. Besides Stafford, New Nine members Young, Conrad, and Lovell would get a chance to command a Gemini flight after being in the second seat on earlier missions.

Perhaps the crash was an ominous sign of sorts, as Gemini 9A was rife with problems during Cernan's EVA, the first since Ed's on Gemini 4. Cernan, a naval aviator and a member of NASA's third group of astronauts, would attempt some specific tasks during his walk, rather than simply floating around, hoping to prove that working in space could be accomplished with ease. How wrong he was. For starters, two launch attempts were aborted—one because the Agena failed to reach orbit, the other for a mechanical issue with the guidance system in the Gemini—then, once in orbit on the third try, the spacecraft could not dock with the unmanned Augmented Target Docking Adapter because the launch shroud failed to deploy once in orbit. Then, on the third day, things got much worse, as Cernan began what he called the "spacewalk from hell." Ed White's only jobs, Cernan noted later, "were to test the spacecraft, the hand-held propulsion device, and the umbilical cord that fed his life-support system from the spacecraft to his suit." For him it had been relatively easy, since he "was such a physical specimen," Cernan wrote, it was "almost like fun." But Gene Cernan would have no fun at all.[79]

Attempting to do some simple tasks outside the spacecraft seemed like it might be easy but every time Cernan tried to turn a valve, for instance, it would cause his entire body to tumble away from the spacecraft until he reached the end of his tether. Each time, he would have to fight his way back to the Gemini and start all over again. He simply could not control his movements. What's worse, he was also fighting the tether, which "felt like I was wrestling an octopus." As a result of his excessive actions, he was sweating profusely and his spacesuit soon filled with 100 percent humidity, causing the visor of his helmet to completely fog over. He couldn't see a thing. The spacesuit "couldn't absorb all of the carbon dioxide and humidity I was pumping out," he said. His heart rate shot up to 180 beats per minute. He was in serious trouble. But he managed to get back to the spacecraft, which splashed down to earth safely a day later. When doctors weighed him after the flight, they found that Cernan had lost thirteen and a half pounds.[80]

Gemini 10, with John Young and Michael Collins, and Gemini 11, with Pete Conrad and Dick Gordon, fared better with their own spacewalks, but not by much. Collins had a lot of trouble ending his spacewalk. Ed had made a "strenuous objection that it did not require 'all his strength and agility' to get back inside the Gemini," Collins wrote in his autobiography, in regards to a memo he had written in 1964 about EVA.

"In Ed's case, I'm sure it didn't, as he was a superb athlete, strong as a horse, but I was really shocked at the difficulty I had in getting that damned pressurized gas bag wedged far enough down in the right-hand seat to provide enough head clearance to close the hatch." Gordon, like Cernan, also tired out very quickly and had to straddle the Agena to rest and regain his strength, causing Conrad to exclaim, "Ride 'em cowboy!" But spacewalking was still more like space-stumbling. With only one mission left, NASA had one last shot at getting it right. And lucky for them, the right guy was in the right spot at the right time.[81]

With the original flight schedule, Conrad and Gordon were to fly Gemini 12, but with the move up in rotation, the last Gemini mission would fall to Jim Lovell and Buzz Aldrin. Aldrin was one of the smartest astronauts in the corps, gaining a doctorate from MIT in aeronautics. In fact, he had written his dissertation on rendezvous, earning him the nickname, "Dr. Rendezvous." He was also a master scuba diver, experience that led Aldrin to begin "neutral buoyancy" training in a deep pool to better simulate the conditions in zero gravity. "It seemed to me that practicing underwater was better preparation for an astronaut's weightless EVA than with the wire-and-pulley training gadgets that came and went in Houston, but never really worked," he wrote. With hard work and very elaborate training, Aldrin "mastered the intricate ballet of weightlessness."[82]

Gemini 12 lifted off on November 11, 1966, for the last mission of what had been a very impressive program—ten flights in less than twenty months. On their way to the pad, the crew made a statement about the final flight of Gemini—Lovell had a sign on his back that read "The" while Aldrin bore one that read "End." Once in orbit, for a four-day mission, it centered almost exclusively on Aldrin and mastering the art of walking and working in space. And it was enormously effective. In three separate EVA's, Aldrin spent a record-setting five-and-a-half hours outside the spacecraft. He utilized foot restraints, waist tethers, and hand-holds to keep him secure and in position while he performed various jobs on the spacecraft and its adjoined Agena. "All the tasks I had been assigned to perform during my space walks had gone smoothly," Aldrin wrote later, "and I had not become overheated or exhausted." It was a flawless performance that accomplished the last objective needed to reach the moon in Apollo.[83]

Gemini had been extremely successful. Despite the scare on Gemini 8, and other issues that were ultimately resolved, all ten missions accomplished their goals. The program had also put America well out in front of the Soviets, in skills but also in the number of missions. By the end of 1966, the Russians had flown only eight manned missions; the United States had doubled that with sixteen. The Soviets were quiet, out of gimmicks and trick plays. But the Americans were unconcerned with their counterparts on the other side of the world. It was now time to move on to Apollo and the final push to the moon. Gus and Ed, who had been instrumental in the success of Gemini, would take a leading role in getting Apollo off the ground. Both of them seemed destined to walk on the lunar surface. What no one knew at the time was, Gemini would be the last time either of them would ever fly into space.

III

THE GOAL: APOLLO

Navy Lieutenant Roger Chaffee sat strapped into his A3D Skywarrior reconnaissance plane ready to fly over hostile territory. It was the beginning of the 1960s, and as the space race was moving forward with greater intensity, the Cold War was threatening to turn red hot. The American–Soviet standoff had always been tense, but now, like in the moon race, the competitors were accelerating at a rapid pace. There were a number of hot spots where nuclear war might erupt at any moment, and citizens in both countries remained on edge. One of the hottest spots was the island nation of Cuba, then in the hands of a communist dictator, Fidel Castro, who had overthrown the U.S.-backed government of Fulgencio Batista in 1959. And once Castro cast his lot with Khrushchev, the game changed significantly. Fear soon spread in the American government, as well as the heartland, that Cuba might become a base of operations for the Soviets. It was, therefore, of the greatest importance that the U.S. military keep tabs on the goings-on in what JFK called an "imprisoned island" just ninety miles off the coast of Florida. To keep an eye on Castro, Air Force and Navy aviators flew reconnaissance missions on an almost constant basis. And one of the newest pilots chosen for flights over Cuba was a young flyer from Grand Rapids, Michigan.

Contrary to some erroneous documentaries on the space program, Roger Chaffee did not fly the vaunted U-2 spy plane in his missions over Cuba. The U-2 was an Air Force plane and Roger was a naval aviator, and a good one at that. In fact, he was one of the youngest pilots in his unit, the Heavy Photographic Squadron 62. Despite his youth, he was called upon to fly numerous missions over Cuba from 1960 to 1962 in the A3D, a twin-engine reconnaissance plane. Because of the advanced state of the aircraft, the Navy only allowed its most experienced pilots to fly it. But Roger learned the plane so well that the Navy gave him permission to take the controls. One of the youngest to do so, he flew it often, sometimes logging as many as three flights a day over Cuba at the height of the crisis. The flights originated from bases on the U.S. mainland or the American base at Guantanamo Bay and involved taking important photographs that proved the Soviet military buildup, including nuclear missiles. Some of those photographs were shown to President Kennedy. When not photographing enemy activity, the unit engaged in other tasks. In one, Roger took photos of Cape Canaveral and its missile launching complex, the very place where his life would change so dramatically before the end of the decade.[1]

Although a young officer while flying over Cuba, Roger was very capable and knowledgeable and could also be quite demanding at times, insisting on perfection. And that's because flying was one of the most dangerous occupations in the world, where pilot error could be fatal. "There's only room for one mistake," Roger would say. "You can buy the farm only once." But despite his demanding ways he had the respect of the men in his squad. "He's tough and an 'eager beaver,'" said one of his colleagues, "but he doesn't ask you to do anything he wouldn't do himself—and we sure love him."[2]

Much like his Apollo 1 crewmate Ed White, Roger was born into flying. His father, Don Chaffee, worked for U.S. Army Ordinance, eventually becoming chief inspector, but he was also a pilot, barnstorming in a Waco 10 biplane at local fairs and other events. He also earned money transporting passengers and flying for parachutists. This allowed Roger to take his first airplane ride at age seven. From that thrilling moment, he was hooked and began building model airplanes with his father, dreaming of becoming a pilot. One day at the age of nine while playing marbles with his friends, a plane flew across the sky. "I'll be up there flying in one of those someday," he told them. Roger liked to play games with other kids but "airplanes were his first love."[3]

Growing up, Roger was "very persistent and [didn't] easily forget little things," even as a toddler. He displayed a high level of intelligence and possessed a wide variety of interests. In addition to building model airplanes, he enjoyed electric trains, which would eventually lead to an avid interest in electronics and electrical engineering in general, shooting and hunting with his grandfather, music—he played three instruments and started a band in high school that was good enough to be hired to play at dances—and the Boy Scouts, eventually achieving the rank of Eagle Scout. Roger also worked to earn spending money. He had chores at home, and, like Gus, an early morning paper route, and small jobs like running errands for the elderly and painting house numbers on sidewalk steps for homeowners, for which he was paid one dollar. He made good grades in school, and aptitude tests revealed that his mind was geared toward mechanics and the arts. He also studied chemistry and thought about becoming a nuclear physicist. He graduated from Central High School in Grand Rapids in 1953 in the top fifth of his class. Another famous alum of Central High was Gerald Ford.[4]

With his stellar achievements in high school, Roger earned a prestigious appointment to the U.S. Naval Academy in Annapolis, Maryland, which he turned down only because he was not quite sure which direction he wanted to go and was not ready to commit to the Navy for an extended period, at least not yet. He also applied for scholarships—a Rhodes scholarship, which he did not get because engineering students were not eligible, and one sponsored by the Naval Reserve Officers Training Corps (NROTC), which he accepted. He began college in Chicago at the Illinois Institute of Technology. While there he joined the Phi Kappa Sigma fraternity and lived at the frat house. During his freshman year at the Institute, he made the Dean's List with a heavy course load of math and science, which he would need to gain a degree in aeronautical engineering.[5]

After completing his first year of college, Roger decided he wanted to pursue a professional flying career in the military and decided to stay with the Navy and transfer to a school with a first-class engineering program. Like Gus, he chose Purdue University in West Lafayette, Indiana, which produced a number of early American astronauts. Neil Armstrong, the first man on the moon, and Gene Cernan, the last man on the moon, were also Purdue graduates. Roger's request to transfer was granted by the Institute, and Purdue accepted him to begin his sophomore year in the fall of 1954.[6]

In the meantime, in order to fulfil his NROTC training obligations, Roger had to spend eight weeks at sea aboard the USS *Wisconsin* battleship. While on the cruise, he visited England, Scotland, France, and Cuba. When he returned home to pass the time before school began at Purdue, he needed a job to earn some money, so his dad got him a job working as a gear cutter at Gear Research. It was hard work and "each day he came out at the end of his looking like a coal miner," something Roger hated about the work. He liked things clean and orderly. But he did come away from the work with a valuable life lesson: "I learned one thing on this job," he said. "I'm not going to make my living *this* way for the rest of my life!"[7]

So, when it was time to head back to school, Roger was more than eager. He arrived on the Purdue campus that fall and secured a room at his fraternity house, where he would live for the duration of his studies. As during his youth, Roger always kept a job to make money. During the three remaining years he was at Purdue, he held three different occupations, some better-paid than others, but all three were better than working in a gear shop. At the beginning of his sophomore year, he waited tables at one of the women's residence halls, then found a job as a draftsman for a local business in town. In his junior year, he was hired to teach freshman math courses. Things were certainly looking up but they were about to get even better.[8]

As his junior year was getting started, in September 1955, lightning struck for Roger Chaffee. He went on a blind date with "a very pretty, brown haired, brown eyed girl from Oklahoma," an incoming freshman named Martha Horn, whom author Lily Koppel described as "drop-dead gorgeous," a girl who "resembled the model 'Twiggy.'" She was "intelligent and capable," "a good listener," and "very feminine." They dated throughout the semester and by January of 1956, she was wearing Roger's fraternity pin, a symbol that she was taken. The Chaffees visited Roger at Purdue in the fall of that year and met her for the first time, when Martha was voted Homecoming Queen. Roger was smitten. "Dad, I've gone out with a lot of girls," he told his father during the visit, "but this is *it*. Some day I'll marry Martha." He popped the question in October, and she happily accepted.[9]

Roger spent the Christmas holidays in Grand Rapids, while Martha went home to Oklahoma City to begin planning for a wedding the following summer. And during the holiday season of 1956, Roger demonstrated the upstanding character that he was known for, even at such a young age. Having been an Eagle Scout, the highest rank in the Boy Scouts, Roger decided to give away all his scouting gear—uniform, sleeping bag, flashlight, knife, and utensil kit—to a needy young scout. So he called his old troop leader to find out if a poor kid in the group was in need of his gear. The scoutmaster knew just who needed it, a young boy who had lost his father and walked across town to attend every meeting, rain, shine, cold, or blazing hot. Roger wrapped up the equipment in nice packaging and he and the scoutmaster took the gifts to the boy's widowed mother, who cried over the kind gesture, for it was the only thing the boy would get on Christmas Day. Unlike today, when everyone seems to want to showcase their acts of kindness for all to see, Roger never told anyone the name of the young scout he gave his prized scouting gear to, which would only have served to embarrass the kid and would have been seen as boasting, which was never in Roger Chaffee's blood.[10]

The spring semester of 1957, his final one before graduation, was jammed-packed. Roger had to complete his degree requirements for a bachelor's in aeronautical engineering and teach his freshman math class, and, on top of all that, he began flight training on a single-engine Cessna 172. He passed his solo flight and received his private pilot's license on May 24, 1957. One of his flight instructors, David Kress, recommended him for military flight training. A little more than a week later, Roger graduated from Purdue. He also received a key to the National Society of Engineers because of his exceptional academic standing. But his educational career was far from over. "It took me four years to learn how little I knew," he said. "Knowledge is vast. There is so much more to learn, and I am going to take advantage of every opportunity that comes along."[11]

During the summer, Roger completed the Marine training that he lacked as part of his naval curriculum, and on August 22, 1957, he received his commission as an ensign in the United States Navy, then promptly drove to Oklahoma to marry Martha. The new couple spent a two-week honeymoon in Colorado before Ensign Chaffee had to report to Norfolk, Virginia. Soon after, in November, he was sent to Pensacola, Florida, to begin flight training, where he flew the T-34 and the T-28, both trainer aircraft. Mastering those beginner jets, Roger was then sent to Texas to learn to fly the F9F Cougar. By November 1958, he was scheduled to begin training on an aircraft carrier, which would require him to leave Martha behind. On November 17, 1958, the day before he was to head out, Martha gave birth to their first child, Sheryl Lyn.

Training on an aircraft carrier is perhaps the most difficult task for any naval aviator. It was no different for Roger. Putting "that big bird down on the flight deck was like landing on a postage stamp," he said, while taking off and landing at night was like "getting shot into a bottle of ink." Yet he mastered it and earned his naval aviator wings in early 1959. Martha proudly pinned them upon his chest at the graduation ceremony.[12]

Over the next three years, Roger's naval career advanced, including his time with the reconnaissance unit. The couple also welcomed son

Stephen in 1961. By the fall of 1962, Roger had amassed over 1,800 hours of flight time in the Navy and had finished his sea duty aboard an aircraft carrier. The Navy offered him the chance to work on his master's degree at Wright-Patterson Air Force Base in Dayton, Ohio, where both Gus and Ed were also stationed for a time. This was the Air Force Institute of Technology, where Navy pilots also studied. Roger's stint there would begin on March 4, 1963, so the family would have to move from their home in Jacksonville to Dayton.[13]

In the summer of 1963, soon after the end of the Mercury Program, Deke Slayton, as the head of Flight Crew Operations, realized that with the upcoming Gemini missions, as well as Apollo, which NASA was deep into planning, he would still be short the requisite number of astronauts needed to fulfil every mission, even with the addition of the New Nine group, which hadn't been in the program a year at this point. So, in June, NASA announced that it would be adding a third group of astronauts, making a change in the requirements. Candidates no longer had to be test pilots. This swelled the pool of candidates to more than seven hundred airmen. Yet NASA wanted no more than fifteen.

This was great news for Roger Chaffee, as he, like Ed, had dreamed of taking his flying career into space. "Ever since the first seven Mercury astronauts were named, I've been keeping my studies up," he said. "At the end of each year, the Navy asks its officers what type of duty they would aspire to. Each year I indicated I wanted to train as a test pilot for astronaut status." Now he would not need test pilot credentials. He had the requisite number of flight hours, which had been lowered to 1,000, and he was of age, although he would be one of the youngest in the program.[14]

Like his first and second group colleagues, Roger had to deal with the same litany of mental tests and physical examinations, and they had gotten no better with time. "They managed to thoroughly humiliate us at least three times a day," he told Martha. After the exams he waited, continuing his work at Wright-Patterson. In October, with a break in his classes, Roger and a friend took a few days off to enjoy deer hunting in

Michigan. He returned to find a message to call NASA in Houston. Without hesitation, Roger returned the call to Slayton and found out he had been selected to join the astronaut corps.[15]

He was elated and immediately called his father. "Dad, I'm in!"

Roger's family was overjoyed by the news, even though he was entering a much more dangerous occupation than his present job in the Navy. "Naturally, his mother and I will worry a bit," Don Chaffee told the press. "But our hearts are content, knowing that our son is doing exactly what he wants to do."

Despite the worry, his folks had confidence that he would be a very successful astronaut. "We knew he'd make it. There wasn't a minute of doubt in our minds. Roger always wanted to be the best. There was only one way for him, the perfect way; nothing less would do." It was a trait they had noticed in him since childhood—a relentless drive for perfection. "When I was a boy, I knew some day men would be going into space," he told a reporter soon after being selected, "and I wanted to go. I've made the goal I aimed for, but now I want to be the first one on the moon, if I can."[16]

On October 14, 1963, more than a year and a half before the first Gemini mission, NASA officially announced the selection of its third group of astronauts, consisting of fourteen new members. Some of these new flyers would fly in Gemini, while others began working on Apollo. Roger was joining an impressive group of pilots. A number of the new astronauts, nicknamed "The Fourteen," would become household names: Michael Collins, Gene Cernan, Bill Anders, Buzz Aldrin, Dick Gordon, Dave Scott, and Al Bean, all of whom would make trips to the moon.[17]

Among Roger's best friends in the program was his fellow third group astronaut Gene Cernan. Roger had been a year behind Cernan at the Naval ROTC program at Purdue. "I really didn't know him during our early years as naval aviators," Cernan wrote in his memoir, but they linked back up again when both were selected as astronauts, arriving together in Houston in January 1964. They became very close. "Roger was my next-door neighbor and one of my closest buddies,"

Cernan wrote. "From the day we reported to NASA, our space careers grew in parallel paths. We shared rental cars, hotel rooms, and often the same airplane." In fact, when Cernan was on Gemini 9, Roger frequently went next-door to be with Barbara Cernan throughout the mission. He was there to answer any question she might have during the flight, which was especially important when things went south during Cernan's EVA. The two astronauts also spent time together socially. "Roger was a workaholic," Cernan said, "but off-duty, he had a great sense of humor." The two often went hunting together to unwind from their intense training.[18]

Walt Cunningham, who would be on the backup crew for Apollo 1, also knew Roger well and had a high opinion of him. "In the early days, some tended to underestimate Roger, perhaps because of his small stature. But he had the capacity to fill a room—any room. It was impossible to attend a meeting with Roger and not be aware of his presence. He had a fighter pilot attitude, even though [his] brief career was in multi-engine photo-reconnaissance aircraft," he said. Like Cernan, Cunningham knew of Roger's intense work ethic and problem-solving skills. "Roger would bore right in—even if it was totally outside his expertise. One of the youngest of the third group, he was fearless, confident, bright, with the all-American-boy look and a beautiful wife to boot."[19]

■   ■   ■

As 1966 began, Slayton began thinking about the crew for the first manned Apollo mission, which could fly as early as the end of the year. In January, a preliminary crew was quietly put together, with Gus Grissom as commander, a deeply satisfying mark of professional distinction for the two-time astronaut. Gus had grown restless in the space program and was entertaining thoughts of flying missions in Vietnam. Though his wife and friends urged him to reconsider, a leadership role in the Apollo program was the tonic needed to get Gus focused on space again. He would never look back.

On March 21, 1966, a few days after the flight of Gemini 8, NASA publicly named the full crew for the first Apollo mission, temporarily called AS-204, or Apollo-Saturn 204. It would be assigned an official mission number as the launch date drew nearer and the unmanned test flights were completed. The prime crew would have Gus as leader with Ed White as senior pilot and Roger Chaffee as the pilot. Gus was very pleased with the assignments. "I think we have a good crew, and I think it will be a good flight," he said. "Ed White's a real hard driver. I don't care what kind of a job you give Ed, he's going to get it done. He's going to get it finished," he said of his senior pilot. "And Roger—of course Roger hasn't had any experience flying [in space] and we don't what he'll be like. I'm sure he'll be okay. Roger is one of the smartest boys I've ever run into. He's just a damn good engineer. There's no other way to explain it. When he starts talking to engineers about their systems, he can just tear those damn guys apart. I've never seen one like him. He's really a great boy." Gus also liked the fact that Roger was known around NASA as "a real hard-ass."[20]

"We'll all be looking forward to the flight," Ed said. And he only had praise for his new commander Gus. "He's good. You know it's interesting how you can really get to know people and understand just exactly how they think and it is quite rewarding when you find out you both think quite a lot alike. This happens to be the case here."[21]

Poised to be the youngest American in space, Roger was overjoyed. "I'm extremely pleased to be named. I think it will be a lot of fun."[22]

Cernan was also pleased when his friend was selected for Gus's crew. Roger "had so impressed our bosses that they assigned him a coveted spot on the first Apollo," Cernan wrote. Cernan himself would be on the support crew for Apollo 1, in the same third-seat position as Roger. Cernan knew Roger's quality, that he worked hard, had little patience for slop, and had the traits to make a top-notch astronaut. This was especially true when it came to the possibility of getting sick in space. A few astronauts succumbed to what became known as "space sickness." Even though they were all very skilled flyers, the sensation of weightlessness caused some to experience nausea, while others saw worse digestive symptoms. "I knew Roger wouldn't," Cernan attested, "because he had an iron stomach that let him eat a banana-sized jalapeno pepper in two big bites."[23]

■   ■   ■

Project Apollo was really starting to get off the ground in 1966, but it is important to understand that it wasn't thrown together at the last minute; it had been around, at least in the idea phase, since Eisenhower. Although Kennedy had made a lunar landing by the end of the 1960s national policy, the concept was birthed in 1959–1960 when initial contracts were awarded to three firms to study the feasibility of a manned lunar landing. NASA even gave the potential new program its name: Apollo, the Greek god who drove his chariot across the Sun. Eisenhower, though, remained skeptical about continued manned spaceflight. In his final message to Congress on the budget, January 16, 1961, four days before Kennedy's inauguration, Ike wrote, "Further testing and experimentation will be necessary to establish whether there are any valid scientific reasons for extending manned spaceflight beyond the Mercury program." Eisenhower's top science advisor, James R. Killian, who had been the president of MIT and chairman of the president's Science Advisory Committee, believed that most "thoughtful citizens are convinced that the really exciting discoveries in space can be realized better by instruments than by man."[24]

When he came into office, Kennedy was looking beyond Mercury for "dramatic results," but his science advisor, Jerome Wiesner, was as skeptical as Eisenhower and his team. In a report on space policy, Wiesner wrote, "*We should stop advertising Mercury as our major objective in space activities,*" he emphasized. "Indeed, we should make an effort to diminish the significance of this program to its proper proportion before the public, both at home and abroad. We should find effective means to make people appreciate the cultural, public service, and military importance of space activities other than space travel." Like the Eisenhower administration, Wiesner wanted to see the Kennedy government embrace unmanned space probes and place an emphasis on science over military matters.[25]

But Kennedy's view prevailed: Apollo would move forward and come to fruition under Lyndon Johnson, who saw no reason to end the program and many reasons to keep it going. Soon NASA began laying out the plan for the push to the moon, as development and production was in full swing on both the new spacecraft needed for the mission and the large boosters that were essential for sending it into orbit and eventually to the moon.

As 1961 drew to a close, before America had even orbited the earth, NASA had completed the awarding of all the major prime contracts for the Apollo program. After a competition with five companies, the contract to build the new Apollo spacecraft went to North American Aviation, headquartered in Downey, California. North American also received the contract to build the second stage of the moon rocket, the Saturn V. Boeing would build the Saturn V's first stage, while Douglas Aircraft won the contract for the third stage. Grumman would design and build the lunar landing craft. The Massachusetts Institute of Technology would construct the guidance and navigation systems for both spacecraft—the command module and the lunar lander.[26]

By the time the major firms awarded contracts to various subcontractors, a total of 47 states had a piece of the moon pie, while nearly half a million Americans worked in some capacity, either directly or indirectly, for the race to the lunar surface in the employ of 20,000 contractors and subcontractors. It was "a kind of middle-class public works program," as NASA bragged about awarding nearly half of all contracts to small independent firms, rather than the big corporations. The program was so large that NASA was "the world's largest organization not dedicated to war," wrote Harry Hurt.[27]

For Apollo to succeed, NASA would need far more powerful boosters, rockets capable of lifting a fifty-ton, three-man crew capsule and service module, and a lunar landing craft for those missions that would travel to the moon's surface, into space and toward the moon. Every booster in Mercury and Gemini—Redstone, Atlas, and Titan—were modified military missiles that were originally used for the sole purpose of war, specifically carrying nuclear payloads, "which were a bit more rugged than astronauts," wrote Deke Slayton. "The boosters vibrated so much during launch that they could rattle an astronaut's teeth, they could damage systems in the spacecraft." So those boosters had to be modified, which took more money and more time to perfect.[28]

In January 1960, with just a year left in office, Eisenhower, though reluctant about spaceflight in general, sent a letter to NASA's administrator, T. Keith Glennan, directing him "to accelerate the super booster program." The only manned program that had been conceived at this point was Mercury, with seven astronauts chosen to carry it out. But by developing larger, more powerful rockets, Ike was at least open to the possibility of more beyond Mercury, although at the start of the new decade nothing was on paper at that point, so von Braun began working on a new series of rockets, the super boosters, which he called Saturn. Unlike their predecessors for Mercury and Gemini, these new boosters that would power Apollo were designed and built from the ground up exclusively for space travel. They were not converted from military systems. The first Saturn, which would eventually become known as the Saturn IB when fully developed for manned flight, was the smaller of the two but it still packed a mighty punch compared to what had come before it. Beginning development in August 1958, the new Saturn I was more powerful than anything America had ever flown, with up to 1.6 million pounds of thrust, and was stronger than anything the Russians had tested to this point. It was designed to boost the Apollo spacecraft into earth orbit but not beyond it, and since the first few missions in Apollo were initially slated to be in earth orbit, the Saturn I was the perfect vehicle. By August 1966, it had been tested thirteen times and had been successful on every flight.[29]

As impressive as the Saturn 1B was, it would not be able to send men to the moon. To do that, von Braun was constructing an even larger, more powerful booster, one that dwarfed the 1B—the Saturn V. It was simply a behemoth in both size and cost. The price tag was more than $100 million apiece. A three-stage rocket, it stood 363 feet in height, which is longer than a football field including both end zones, was 33 feet in circumference at its base, and weighed 6.5 million pounds fully fueled. It was roughly the size of a U.S. Navy destroyer. The Saturn V had 3 million parts and could generate a tremendous amount of power, an astounding 7.6 million pounds of thrust, equivalent to 160 million horsepower, twice as much power as every single river and stream in the United States running through a hydroelectric power plant simultaneously. In fact, the "output of the Saturn first stage was 60 gigawatts," similar to the peak electricity demand of the United Kingdom. Another calculation placed its energy equivalency at 85 Hoover Dams. To generate such an awesome amount of power, it used nearly a million gallons of several different types of propellants, both a fuel and an oxidizer, which "enables it to burn in airless space"—kerosene, liquid oxygen (at -297 degrees Fahrenheit), and liquid hydrogen (at -423 degrees F).[30]

Each of the Saturn V's first stage F-1 engines, which to this day are the most powerful engines ever built, were 18 feet in diameter, burned 6,000 pounds of propellant every second—15 tons for all five engines—pushed by fuel-pump turbofans churning at 20,000 revolutions per minute and pumping in 42,500 gallons of propellant, the force of thirty diesel locomotives. All five F-1 engines, working together, could drain an entire Olympic-size swimming pool in 30 seconds. To build up enough thrust to lift off the pad, ignition sequence start—the lighting of the engines—occurred at T-minus 8.9 seconds. On liftoff, as the massive rocket began shaking as it built up to full thrust and large chunks of ice from the super-cold fuel began falling from the fuselage to be consumed in the inferno below, the hold-down arms that kept the rocket upright on the pad, had to release simultaneously. If just one hold-down arm released early, even by as little as a split second, the engines had such great force that the entire rocket would topple over and explode. At full thrust and on its journey upward, the engines of the Saturn V's first stage produced enough heat to melt part of the concrete structure into glass and produced enough noise to destroy the rocket, as sound waves would bounce off the ground below and ricochet back upward. So to combat this problem, 50,000 gallons of water per minute flowed through a flame trench under the launch pad to divert the flames and suppress the sound waves. But the noise was still intense, so much so that the first time the engines were tested at the testing facility in Huntsville, the sound waves broke windows 15 miles from the test stand. When the first fully-configured Saturn V was tested in 1967, the sound and vibrations caused problems in Walter Cronkite's viewing booth several miles away. "Our building's shaking here, our building's shaking! Oh it's terrific…the building's shaking! This big blast window is shaking! We're holding it with our hands! The roar is terrific!… Part of our roof has come in here!" he exclaimed excitedly while watching the first unmanned Saturn V flight.[31]

Soon after launch, the Saturn V broke the sound barrier—738 mph— after just one minute of flight at an altitude of one mile. At the three-minute mark, the first stage had consumed 5 million pounds of propellant, "enough fuel to fly twenty Boeing 707s from New York to Paris and back." At three minutes and five seconds, the first stage, which had reached an altitude of 43 miles, cut off and was jettisoned, and the second stage ignited. The weight of the machine was down to 1.5 million pounds. The second stage burned out after six minutes and was jettisoned, igniting the third stage, the S-IVB, for a burn of 1.5 minutes, which would place the spacecraft in orbit. Then, after one-and-a-half revolutions of the earth, the S-IVB would ignite a second time and push the spacecraft out of earth orbit and toward the moon. Even though Gus and his crew would fly the smaller Saturn IB rocket to earth orbit on Apollo 1, all three hoped to be commanding a later mission on a Saturn V bound for the moon.[32]

To assemble the Saturn rockets, especially the three stages of the Saturn V, NASA constructed the largest building in the world, the Vehicle

Assembly Building (VAB). In the shape of a cube, it stands 525 feet high and 518 feet wide, covering eight acres, more than enough room for the entire New York Yankees playing field to fit comfortably on the roof. In fact, the Statue of Liberty could fit inside the VAB with 220 feet of space left over. The bay doors are the largest in the world at 456 feet in height and take about 45 minutes to open or close. To transport the rockets from the VAB to the launch pad, NASA built what is known as the "crawler-transporter"—a moving platform 131 feet long, 114 feet wide, powered by two 2,750 HP diesel engines, which consumed 150 gallons of fuel per mile. The rocket, though not yet fueled, still had to remain vertical, so the crawler literally crawled, cruising at less than 1 mph on steel treads that resemble that of a tank, yet far larger and heavier with each tread weighing one ton. By itself, the crawler weighs 6.6 million pounds and can carry 18 million pounds (approximately 20 fully loaded Boeing 777 airplanes). Transporting the Saturn V to the pad on Merritt Island, a distance of 3.5 miles, and assembling it on the launch pad took anywhere from 6 to 8 hours.[33]

Riding on top of a Saturn rocket would be a new Apollo spacecraft, a conical-shaped, three-man capsule, which resembled a gumdrop. It was called the Command Module and was attached to a Service Module, which was cylindrical-shaped and carried the fuel cells, oxygen tanks, and the propulsion system, the SPS engine. The entire apparatus would weigh 100,000 pounds and was built by North American Aviation. Like its Gemini predecessor, the Apollo command module would have a guidance computer that was top of the line for its day but primitive by today's standards of computing. In fact, an Apple iPhone with 16 gigabytes of memory is 8 million times more powerful than Apollo's guidance computer and even has more computing power than every supercomputer at NASA's Mission Control Center in Houston. All the guidance computer could do was keep the spacecraft pointed in the right direction and on track to and from the moon; it could not compute the trajectory. That had to be done by flight controllers in Houston. But for its day, Apollo was a very impressive and complex machine.[34]

The Soviets got wind of the new concept and decided they needed to get ahead of American innovation. In October 1964, the Russians launched a Voskhod spacecraft, presumably a new, three-crewed capsule. But it was yet another Soviet publicity scheme. NASA found out later that it was nothing more than a Vostok spacecraft, the Russian two-man capsule that they had been using, with three smaller seats in place of the two-seat system. The three cosmonauts, crammed in a space for two men, were sent into space for one day with no helmets or spacesuits and no launch escape system, to save on weight. "I guess the idea was to score another public relations first," said Deke Slayton.[35]

The concept NASA had chosen to fly to the moon demanded taking a three-person crew. There were a number of possible avenues to land on the moon—Direct Ascent, Earth Orbit Rendezvous (EOR), or Lunar Orbit Rendezvous (LOR), and NASA was split on which approach would work best. Direct Ascent was the simplest concept but would be the hardest to pull off, and would most likely take a lot longer to accomplish, not to mention a much higher cost. It consisted of crafting a massive rocket, at least twice as big as the Saturn V, a proposal known as Nova, and flinging the entire rocket into lunar orbit. Once there it would use braking speed to slow down and land on the surface of the moon, which might be a problem since it would be about 80 to 100 feet tall. One engineer said it would be like trying to land the Washington Monument on the moon's surface. To return, the rocket would blast off the lunar surface and back to earth. By some estimates, that feat of engineering could take up to ten years longer to accomplish, especially given the fact that the Saturn V itself was still years away from completion at that point. Nova would take even longer to develop.[36]

EOR, or earth orbit rendezvous, which von Braun favored, required multiple Saturn launches, each one carrying separate components that would be assembled in earth orbit, then flown to the moon. But that also had major problems, such as the exact timing of each launch so rendezvous could be accomplished for each piece, then the complexities of linking everything together. It seemed to be far too complicated. But

NASA engineers, led by Dr. John Houbolt, came up with a new approach, lunar orbit rendezvous, or LOR. It would take one booster, the Saturn V, to put both spacecraft in orbit, then the third stage would be fired a second time to send the spacecraft toward the moon. Soon after, the third stage would be jettisoned and the command/service module would dock with the lunar module while en route to the moon. Once in lunar orbit, two crew members would board the lunar lander, which would detach and descend to the surface, leaving the command/service module with the command module pilot in lunar orbit, then, after blasting off from the surface, the lunar module would rendezvous with the command module back in lunar orbit. This seemed a much more likely option. After studying the question for some time, and arguing about it to such a degree that *Time* called it the "Moon Spat," NASA ultimately chose LOR and announced it publicly on July 11, 1962. All that was left now was to build the components.[37]

North American Aviation won the contract in November 1961 but did not produce a completed command module until 1966. One reason was the complexity of the spacecraft, which was, up to that time, the most complicated machine ever devised. Deke Slayton referred to Apollo as "horrendously complex." At 36 feet long, the craft was far larger than Mercury and Gemini and, with the service module attached and fully fueled, weighed five tons. The electrical system comprised 30 miles of wiring with 13,000 electrical segments. The control panel in the cockpit portion of the command module consisted of 640 switches, circuit breakers, event indicators, and computers. Yet despite its increased size over the Gemini spacecraft, the Apollo command module was still a tight fit for three astronauts. "It was like you had squeezed three guys in the front seat of a Volkswagen Beetle," said Apollo astronaut Charlie Duke. Buzz Aldrin said it was "hardly claustrophobic" in space but, on the ground, it was "definitely cramped" because all three astronauts were in their bulky spacesuits and helmets. In space, without the suits on, it was much roomier. The capsule had a lower equipment bay with an area for astronauts to get some sleep, unlike its simpler predecessors, so the crew could actually get out of their seats and move around while in orbit.[38]

With planning and development beginning in 1961 and changes coming at a rapid pace, North American ended up with two versions of the same spacecraft, a Block I model and a Block II. The Block I's were, in essence, a "trial model" without some of the features that would be included in a Block II. One conspicuous absence was the docking apparatus. The Block I could not dock with a lunar lander, nor did it have the pressurized tunnel that allowed the crew to move between the command module and the lunar lander, which hardly mattered since the Block I's were going to be used in earth orbit–only test flights, like Apollo 1. The Block I also did not have a quick-release, outside opening hatch, but boasted an inner hatch that opened to the inside, and an outer hatch that opened to the outside. This cumbersome system would play a major role in the Apollo 1 fire. Development continued on both versions of the spacecraft with the first Block I due at the Cape in the summer of 1966.[39]

■   ■   ■

With all the massive building and massive spending running right along with it, NASA began planning the Apollo flight roster to fly the first spacecraft. The initial schedule called for Gus, Ed, and Roger to fly the first manned mission, AS-204, in a Block I command module. AS-201, 202, and 203 would be unmanned tests of the Saturn I rocket and the spacecraft. So when the official missions were named, Gus's mission should have been Apollo 4. But Gus insisted that the first manned flight of the lunar program be Apollo 1, and he even had an official mission patch designed to reflect his desire. NASA, though, would not officially designate it as such until after the fire.

Gus's original backup crew for Apollo 1 would be Jim McDivitt, Ed's old friend and partner on Gemini 4; Dave Scott, who had flown with Neil Armstrong on Gemini 8; and rookie Rusty Schweickart. Gus and his team would fly on a Saturn IB rocket into earth orbit and test the command and service module in a lengthy shakedown cruise, perhaps lasting as long as two weeks. The mission was officially designated as "open-ended," and Gus was given control over how long the flight would continue. He was determined to see it last a full two weeks in order to fully put the new spacecraft through its paces. The second crew for AS-205 would be Wally Schirra, Donn Eisele, and Walt Cunningham for a repeat performance in a Block I spacecraft in low earth orbit. McDivitt and his crew would then fly aboard the larger Saturn V in the more advanced Block II with a lunar lander and test it in earth orbit. But the development of the lander was considerably behind schedule and before long NASA concluded that the schedule would have to change.

Realizing that two Block I, earth-orbit missions would be redundant and unnecessary and with pressure from Schirra, NASA cancelled AS-205 on December 22, 1966. The crew of Schirra, Eisele, and Cunningham became the backup crew for Apollo 1, while McDivitt, Scott, and Schweickart would fly the second mission with a lunar lander, then under construction by Grumman. Schirra thought his crew might nab that mission but it went to McDivitt instead because he and his team and been training extensively for it. The third Apollo flight would be Frank Borman's crew, which initially included Michael Collins and rookie Bill Anders. It would also fly on a Saturn V to high earth orbit, around four thousand miles, to test the lander and reentry procedures.[40]

Despite the shuffle in flight crews, which would continue to change right up to the first flight, everything seemed to be falling into place, just like every previous mission in Mercury and Gemini. The construction of Spacecraft 012, designated for Apollo 1, began with the initial fabrication in August 1964, and the basic structure was in place by the following September. After the installation of all the subsystems in the command module were completed in March 1966, the spacecraft was delivered to the Cape on August 26, 1966. The following month, it was mated to the service module.[41]

At first, a number of astronauts were impressed with North American Aviation. Their quality was very well known. The company "did have a first-rate facility at the Downey plant," wrote astronaut Al Worden. Every Apollo spacecraft was assembled in a "clean room" that was immaculate, complete with air scrubbers that had the same standards as a surgical operating room in a hospital. To enter the clean room, every person had to don protective clothing including head covering, then walk across a sticky pad to make sure no debris was stuck to the bottom of one's shoes, then pass through an area with large fans to blow any remaining substances, like dust particles, off the protective clothing. Once inside the large room, there were numerous command modules lined up in various stages of completion. "It was like entering a science-fiction movie," Worden said.[42]

Running the Apollo Spacecraft Program for North American Aviation was Harrison "Stormy" Storms, whom Tom Stafford described as "a big, smart, tough individual." North American "had strong political connections in Washington with a great lobbying team. North American was a slick, big-time bunch of Washington operators compared to the mom-and-pop operation at McDonnell," Stafford noted. "Or so we thought," but troubles were beginning to mount in regard to the new spacecraft. "I hadn't made too many trips out there...before I began to see all kinds of problems."[43]

North American's training simulator was also a complete mess. A flight simulator was one of the most important training devices for the flight crews. It had to perfectly mimic the actual spacecraft. It should be no different to a pilot whether he was in the simulator or the actual spacecraft. And the simulator for Apollo did not match the spacecraft. In many ways it wasn't even close.

The first time Frank Borman stepped into the Apollo simulator he got quite a shock when he used the hand controller to maneuver the spacecraft. When he pulled it back, the nose of the spacecraft went down, not up, then he pushed it down and the nose went up, not down. "I called the engineer over and I said, 'You got the polarity reversed on this hand controller,'" Borman said. "And he said, 'Oh no, that's the way we're going to use it. That's the way we're going to fly it because it makes rendezvousing easy. It makes docking easier because, when you pull back on the stick, your nose goes down but the target goes up. You see,' and 'That's the way we were going to do it.' But this is another example of NASA. I said, 'Well, look, that may be the way you're going to do it, sitting here on your ass as an engineer, but that's not the way we're going to do it.' And I called back to the Apollo Program Office, and I got it changed right there."[44]

Gus, though, was dealing with the simulator on an almost daily basis for months and was furious over it, and rightfully so. One day his frustration overflowed. He came in for some simulator time with Ed and Roger and barked at Riley McCafferty, who was in charge of it, right off the bat, "Let's see if this thing will fly, Riley," half-believing his own words.

And, of course, it didn't. "Damn it, Riley! This simulator is worthless!" Gus seethed. "Too many things don't match."

The simulator simply wasn't up to speed. Changes were coming in at an ever-increasing rate for the spacecraft, particularly in terms of software for the computer, and the simulator had not been updated to keep pace.

McCafferty tried to explain to Gus that it took some time to get the simulator updated but Gus was having none of it. "It's a piece of crap, Riley. Get it right and we'll be back!" Gus said. But before he left, he had a little memento he wanted to leave to showcase his displeasure for all to see. Reaching in his briefcase, Gus brought out a lemon and hung it on the simulator. "Leave it there," he exclaimed as he left the building. It was just five days before the all-important "plugs out" test.[45]

The spacecraft didn't look much better. "The Block I spacecraft design was highly defective, pure and simple," said John Young. "When Gus, after one checkout run at Downey, left a lemon on top of the CM simulator, he wasn't really joking." Gemini and Apollo astronaut Jim Lovell called the Block I spacecraft "a disaster."[46]

Chris Kraft, the director of flight operations in Houston, got wind of the problems at North American Aviation. He sent John Bailey, a rocket engineer who had been at NASA from the very beginning, to

Downey to see what was going on. In his report to Kraft he wrote, "This hardware is not very good. The cabling is being stepped on when they work on the spacecraft. There's no protection for it. The people are not very good at checking this thing out. They're not very good at trying to maintain some semblance of the fact that a human being is going to be in this machine. I'm telling you, it's not good." But like every other warning, this one also fell on deaf ears, as many inside NASA figured the problems would be worked out before the first flight.[47]

During the previous summer and fall, when he was having so much trouble with the flight simulator, and North American was having so much trouble with the actual spacecraft, Gus had let his feelings be known to everyone he could—from engineers and technicians to NASA brass. And everyone promised Gus things would get better. "Gus was fussing a little bit," said Stephen Clemmons, who worked as a technician for North American. "Of course Gus always fussed. Sometimes we'd call him 'The Nitpicker.' He probably had a right 'cause he was very unhappy with the spacecraft and he didn't bother to conceal his feelings on it." But Gus felt "like a wolf howling in the wilderness." He even let out his frustrations in a backhanded manner at a press conference in late 1966. When asked what would be considered a successful mission, Gus answered, "As far as we're concerned, it's a success if all three of us get back." The response caused laughter from the assembled reporters but nothing more. There were no probing questions for NASA engineers. As space historian Jeffrey Kluger has written, "the junkyard spaceship got no better."[48]

Gus even vented to NBC reporter Jay Barbree. "Apollo is a piece of crap," he told him. "It may never fly. We have problems and they're not getting solved. It's nothing like Mercury and Gemini and working with the Mac folks in St. Louis. Hell, these California boys in Downey haven't a clue. They've got their big fat contract and no know-how. Apollo is not ready."[49]

He raised similar concerns to Betty, which was unusual, she said, because he never brought work problems home with him. "I knew it wasn't as smooth going as it had been in the other flights, getting the capsule ready. He got a number of phone calls at home and they would be fussing about something. That was not like Gus." He didn't like to bring the space program home.

"If there was a problem in the Mercury or Gemini capsules," Gus told Betty one day, "and if I couldn't get the regular people to listen or try to fix it up so it was workable, I could go to Mr. Mac," who was the president of the McDonnell Aircraft Company, "and get some kind of satisfaction. It's not that way with North American."

"How come?" she asked. "Well, there are just too many bosses," he said.[50]

Gemini and Apollo astronaut Frank Borman also saw the differences between McDonnell and North American. "It was like going from night to day. McDonnell was much more informal and, you know, I think everybody really had a great deal of respect for Mr. McDonnell.... And it was a sort of a countrified company. You just had the feeling of people that did their work and weren't very fancy about it. You went to North American, and they had layers and layers of briefers and this and that and the other things, and customer service or customer reps. I don't know. It was also a feeling what—that I had the feeling, 'Well, we built the X-15. We did this. We did that. We know more than everything about it.'"[51]

Gus also felt strongly that one of the major problems was rushed production. He told Barbree that he suspected this may have had something to do with the upcoming presidential election. If Apollo landed on the moon in late 1967 or, better yet, sometime in 1968, the re-election of Lyndon Johnson to a second full term would likely be a cinch, given the ongoing problems mounting around the country. Other astronauts understood the rush too. "There was a lot of pressure within NASA to get off on time," said Walt Cunningham.[52]

Barbree looked into it, at Gus's suggestion, but it didn't do much good. "In the coming days, I questioned Apollo managers often and regularly. I wanted to know why they weren't addressing problems that had been brought to my attention," Barbree wrote. "I wanted to know why they were in such an all-fired hurry to launch in late 1967 or early 1968. John Kennedy had set the launch for before the decade was out. Why didn't they take their time? Was beating the Russians more important than astronauts' lives?" He got few answers and was nearly alone in his complaints, as other reporters just shrugged off his concerns. "The press and public ignored the whole damn thing, and the first Apollo labeled 'flight worthy' was soon sent to the Cape."[53]

As Barbree found out, NASA management, particularly of the Apollo program, was also suspect. "NASA's Apollo program office (ASPO) was making its own mistakes," wrote Tom Stafford. ASPO in Houston was headed by Joe Shea, "a brilliant engineer from the electronics world" who took over in 1963. "Joe was also a protégé of George Mueller's. Unlike Mueller, though, Joe was the same age as most of us in the astronaut office, which made him very competitive. We'd play handball with him, for example. But whenever issues or problems came up, he was no longer Joe Shea, the other guy on the handball court, but Mr. Shea, the program manager. We would become close friends years later, but during Apollo, Joe and I had some real head-knockings."[54]

For his part, Shea blamed North American for most of the problems. He believed that North American was only motivated by money and often butted heads with managers in Downey. He and Storms did not like each other at all. Shea did not have "a high opinion of North American and their motives in the early days," he said years later. "I think they were more interested in the financial aspects of the program than in the technical content of the program. I think Storms was a very bad general manager, I think Atwood [North American's president] had dollar signs in his eyes. Their first program manager was a first-class jerk. There were spots of good guys, but it was just an ineffective organization. They had no discipline, no concept of change control. If anything, they were interested in pumping the program up rather than in what the program really was."[55]

Despite the conflicts, when Spacecraft 012 was delivered to Kennedy Space Center in August 1966, North American issued a Certificate of

Worthiness, meaning it was considered flightworthy. Yet along with the certification was a list of engineering issues that had not been resolved, some 113 individual projects. "It was not finished. It was what they called a lot of uncompleted work or incomplete tests," said Wally Schirra. "So it was shipped to the Cape with a bunch of spare parts and things to finish out." During a "combined systems" test of the command and service module, which ran from September 14 to October 1, there were 152 inconsistencies with various systems, with one major problem being a short in the radio command system. Yet on October 7, a board chaired by Dr. George Mueller, the associate administrator for Manned Space Flight, certified the vehicle "flightworthy, pending satisfactory resolution of listed open items."[56]

In mid-October, an unmanned altitude chamber test was completed with a grade of "satisfactory." Three days later, during a manned test in the altitude chamber, a problem occurred at 4,000 feet when a transistor failed in one of the inverters but after it was changed out the test was concluded with similar satisfactory results. But on October 19, during a repeat of the manned altitude chamber test, another problem cropped up. The major trouble now was the environmental control unit, and it had some serious flaws. A problem with the oxygen cabin supply regulator that North American couldn't solve was discovered during the test, leading to an indefinite suspension of another manned test later that month. On October 27, technicians removed the entire environmental control unit and sent it back to the factory in California for not only a repair but a change in design. It wouldn't be the last problem with that particular unit. Subsequent units leaked glycol, the coolant to keep the cabin comfortable, onto the floor of the spacecraft. By mid-December, three different environmental control units had been put into Spacecraft 012, which was to be flown by Gus, Ed, and Roger.[57]

As technicians were literally working around the clock to get Spacecraft 012 at the Cape up to snuff, there were other issues with North American's production of additional command modules in California. A propellant tank in the service module of Spacecraft 017 burst. Making matters worse, the rupture did not come during a pressure test, meaning it was not under any undue pressure. Therefore, the propellant tanks in the 012 service module had to be checked out to make sure they were flightworthy.[58]

As the new year dawned, it looked as if things just might be getting on track. During the final manned altitude chamber test, every spacecraft system functioned normally. The spacecraft was removed from the altitude chamber on January 3, 1967, and mated to its Saturn IB booster on January 6. It was then taken by the crawler transporter to Pad 34 for one of the last major tests on January 27 before the scheduled launch date of Apollo 1 on February 21.[59]

By now, the crew was becoming more comfortable with the vehicle that would take them into space. Even the crew's fresh-faced rookie Roger had expressed his supreme confidence in the new Apollo spacecraft. "I think we've got an excellent spacecraft. I've lived and slept in it. We know it. We know that spacecraft as well as we know our homes, you might say. Boy, we know every little rivet and wire and electrical termination in it practically. We're confident it's a darned good spacecraft."[60]

In fact, the entire Apollo 1 crew "felt that the defects that had been noted throughout the development had been corrected and the spacecraft as it existed prior to this test was believed to be in good shape," Frank Borman would later say. In the official report on the fire, issued from the committee on which Borman served, it noted that the final manned test of the command module, in an altitude chamber with the back-up crew, "was very successful with all spacecraft systems functioning normally." The report further stated that the "system was determined to be ready for the initiation of the Plugs-Out Test on January 27, 1967."[61]

But Gus, Ed, and Roger did still have concerns. As a joke, though with a lot of truth to it, the crew sent a signed picture to Joe Shea. In the photo, Gus, Ed, and Roger were bowing their heads, with eyes closed and their hands clasped together in prayer before a model of the command module. Their inscription read: "It isn't that we don't trust you, Joe, but this time we've decided to go over your head."

■   ■   ■

As Apollo 1 was moving closer to liftoff, there was more trouble brewing, not only with technical issues but genuine opposition to the program, as hostility was growing across the country. The 1960s had opened with such hope and promise for a brighter future than the image of the stuffy old 50s. A new generation of Americans, John F. Kennedy reminded the nation in his inaugural address in January 1961, was now at the forefront of national leadership. Camelot had become the symbol of a young, energetic people on the move, and space exploration would be a big part of it.

But by the mid-60s that dream seemed to be fading a bit. Kennedy was dead and the new president, Lyndon Johnson, a throwback to the politicians of old, continued to push for spaceflight and the lunar program. But he also had other ideas about what would make America a great society. It was not a face-off with the Soviets that drove Johnson but a transformation of the American homeland. With an agenda that dwarfed the New Deal, LBJ's Great Society included massive new government programs to tackle poverty and reform education, as well as providing healthcare for the poor and the elderly. The Civil Rights Movement was in full swing, and the country was rife with major demonstrations as well as major violence. Congress, with LBJ's herculean support, passed two major bills in 1964 and 1965 to bring about full racial equality, at least on paper. But it was not Johnson's war on poverty or the fight for inequality that caused a national uproar, but a real war in southeast Asia that threatened to undo the very fabric of the nation.

As Alan Shepard and Deke Slayton wrote in *Moon Shot*, "The year 1967 rolled in like a political garbage truck with its tires burning." As Apollo was set to fly early that year, nearly half a million U.S. troops were fighting in Vietnam, even though President Johnson had promised in 1964 that "American boys" would never be sent ten thousand miles away from home "to do what Asian boys ought to be doing for themselves." Yet three years later, Americans were fighting and dying every day. The country was growing weary, and the streets were filled with those who opposed the draft and wanted the troops and the nation out of the war.[62]

Apollo should have been the unquestioned bright spot in what was shaping up to be a decade of tumult, rather than a decade of triumph. America could still emerge as the hope of the free world with a triumphant landing on the moon, yet many Americans were beginning to question the lunar program and even the manned spacecraft program itself. Perhaps the biggest issue was the astronomical amount of taxpayer cash devoted to Apollo, with estimates ranging from $20 billion to as high as $40 billion, figures that reflected the spending for both Mercury and Gemini. Some found such a massive budgetary expense a waste and thought the money would be best spent elsewhere. Others thought the whole thing was nothing more than a silly, childish stunt.[63]

Many leading citizens, politicians, writers, engineers, and scientists had their doubts. One question dogged Norman Mailer "into the tenderest roots of his brain," he wrote in *Of a Fire on the Moon*: "Was the voyage of Apollo 11 [the lunar landing flight] the noblest expression of a technological age, or the best evidence of its utter insanity?" Famed journalist Walter Lippmann took a few jabs at the program, and the man who set it in motion, writing in 1963, "There were two big mistakes. One was the commitment to put a man, a living person rather than instruments, on the moon," he wrote. "The other mistake was to set a deadline—1970—when the man was to land on the moon. These two mistakes have transformed what is an immensely fascinating scientific experiment into a morbid and vulgar stunt.... For this is showmanship and not science, and it contaminates the whole affair." One sociologist called the entire program a "moondoggle."[64]

Two years after leaving office, former president Eisenhower was still criticizing Apollo. "I have never believed that a spectacular dash to the moon, vastly deepening our debt, is worth the added tax burden it will eventually impose upon our citizens," he said in 1963. Ike was looking at the issue through a conservative lens. Not only should the money not be spent elsewhere, but it should not be spent at all. While still in office, in 1960, when told the price tag for a moon program, Eisenhower resorted to the blunt rhetoric of his military days. "I'm not about to hock my jewels," he said. But not every critic thought about it in such a conservative way.[65]

Others, particularly those on the left of the American political spectrum, thought the money should be shifted to more social programs. "You dig 50 pounds of moon rock and what do you get? Another day older and deeper in debt," said writer Kurt Vonnegut, who was a major skeptic of the space program. "For that kind of money, the least [NASA] can do is discover God," he told CBS News. "We have spent something like $33 billion on space so far. We should have spent it on cleaning up our filthy colonies here on earth." In one notable interview, a reporter asked Vonnegut, "You've lately taken a couple of swipes at the space program." Vonnegut interjected, "I think it would be interesting to talk more about whether this is the proper thing to do with the public treasury. The sort of dreams I would have would be a habitable New York City, for instance. It would seem to me that that would be a reasonable thing to do."[66]

One mathematician, Warren Weaver, a past president of the American Association for the Advancement of Science, calculated what could be purchased with the estimated $30 billion NASA was projected to spend to get to the moon before the end of the decade:

> One could give a 10 per cent raise in salary, over a 10-year period, to every teacher in the United States from kindergarten through universities (about $9.8 billion required); could give $10 million each to 200 of the better smaller colleges ($2 billion); could finance seven-year fellowships (freshman through PhD) at $4,000 per person per year for 50,000 new scientists and engineers ($1.4 billion); could contribute $200 million each toward the creation of 10 new medical schools ($2 billion); could build and largely endow

complete universities with liberal arts, medical, engineering and agricultural faculties for all 53 of the nations which have been added [up to that time] to the United Nations since its original founding ($13.2 billion); could create three more permanent Rockefeller Foundations ($1.5 billion); and one would still have left $100 million for a program of informing the public about science.

Kenneth B. Clark, an African-American psychologist at the City University of New York, testified before Congress about the plight of the inner cities, which would only get worse by going to the moon. "I just don't think the moon is going to be an adequate substitute for the fact that we haven't addressed ourselves to clearing up the slums."[67]

Gil Scott-Heron, a black poet and blues singer, penned lyrics to express his displeasure in sending men, white men at that, to the moon. The song was entitled, "Whitey on the Moon."

> A rat done bit my sister Nell.
> (with Whitey on the moon)
> Her face and arms began to swell.
> (and Whitey's on the moon)
> I can't pay no doctor bill.
> (but Whitey's on the moon)
> Ten years from now I'll be paying still.
> (while Whitey's on the moon)
> The man just upped my rent last night.
> ('cause Whitey's on the moon)
> No hot water, no toilets, no lights.
> (but Whitey's on the moon)
> I wonder why he's upping me?
> ('cause Whitey's on the moon?)
> I wuz already paying him fifty a week.
> (with Whitey on the moon)

Taxes taking my whole damn check,
Junkies making me a nervous wreck,
The price of food is going up,
An' as if all that shit was't enough:
A rat done bit my sister Nell.
(with Whitey on the moon)
Her face and arm began to swell.
(but Whitey's on the moon)
Was all that money I made last year
(for Whitey on the moon?)
How come there ain't no money here?
(Hmm! Whitey's on the moon)
Y'know I just about had my fill
(of Whitey on the moon)
I think I'll send these doctor bills,
Airmail special
(to Whitey on the moon).[68]

There was also opposition in Congress, particularly during discussions of NASA's ever-growing budget requests. Senators like William Proxmire of Wisconsin, who was always proposing cuts to NASA and would eventually get the Saturn V production lines shut down, Joseph S. Clark of Pennsylvania, Walter Mondale of Minnesota, and J. William Fulbright of Arkansas, expressed disapproval of the lunar program and NASA in general, particularly the spending. In 1958, NASA's budget was $189 million, which shot up to $691 million the following year. In 1960, it hit the billion-dollar mark, with an appropriation of $1.031 billion. By 1963, as Mercury was ending, NASA was asking for nearly $6 billion in the budget for Fiscal Year 1964, but the House was the first to balk, cutting $600 million. Kennedy argued that NASA needed at least $5.4 billion to keep Apollo on schedule and turned to the Senate to restore the cuts. There Fulbright introduced an amendment for a further cut of 10 percent, but his effort failed by ten votes. The Senate Appropriations

Committee tried to restore $90 million to NASA but Proxmire was successful in blocking that effort.[69]

Mondale, representing the liberal wing of the Democratic Party, wanted the money spent on social programs. Though he said years later that he supported Apollo, he had more passion for "progress in education, dealing with poverty, and so on at home." Fulbright, who was not necessarily opposed to the moon landing program but had reservations with the compressed time line, also wanted to divert some of NASA's funds to other "pressing problems here on earth," he told the Senate during the debate over his amendment. "The probable truth is that we are in a race not with the Russians but with ourselves," he said just three days before Kennedy's assassination. "Khrushchev's latest statements, which may or may not be taken at face value, indicate that the Russians are continuing their efforts to send a man to the moon, but do not wish to engage in a race with the United States. It may well be that we have entered a trap of our own making, that we have committed ourselves to a futile race of which the outcome can only be outright failure or a pyrrhic victory."[70]

Though true in some respects, Kennedy did offer the Soviets a joint mission to the moon, which Khrushchev turned down. As late as 1966, the *New York Times* was editorializing about the moon race. "It is still too early to predict whether English or Russian will be the native language of the first man on the moon," the paper opined, "but the prospect is that no more than a few months will separate the two nations' separate accomplishment of this historic feat."[71]

Fulbright, though, continually pushed his own viewpoint about NASA's budget and even discussed the issue with President Kennedy in the White House when Congress was considering a large appropriation for NASA. He thought the money would be better spent on education. "Bill, I completely agree with you," JFK said. "But you and I know that Congress would never pass that much money for education. They'll spend it on a space program, and we need those billions of dollars in the economy to create jobs."[72]

The opposition members of Congress, though, continued to voice their disagreement, even as it ran counter to the president. Speaking on the Senate floor the day after Fulbright, Senator Clark quoted the distinguished columnist James Reston, "The question is which comes first—the moon or the slums, the unexplored or the unemployed, security or solvency." But Clark was not seeking to end Apollo or NASA, just to prioritize things. The moon program was "an exciting adventure. It is not only in our national interest, but in the international interest of all mankind. The only problem is, how much, how soon, and at what cost to other programs?"[73]

And the cost was considerable and climbing. From 1960, Dwight Eisenhower's final year in office, to 1965, under Lyndon Johnson, NASA's budget had grown by 900 percent. In any given year, NASA consumed anywhere from 3–5 percent of the overall federal budget. NASA budgets in 1964, 1965, and 1966 would top $5 billion each year, while the 1967 budget came in at $4.97 billion. But steeper cuts were on the horizon, as appropriations in 1968 and 1969 were set to be trimmed by $1 billion overall. Some have used the cuts to conclude that LBJ was never that interested in space, or was at least losing interest in it as time wore on. The truth is that Great Society costs were spiraling, as was the war in Southeast Asia. Senator Walter Mondale concluded that "what broke the bank was the war in Vietnam," as it "kept costing more and more." To keep up, Congress would have to make cuts in a number of different areas, and space was one of them.[74]

Another issue was the science, or lack thereof. A number of scientists believed there would be no tangible benefit from a manned expedition to the moon. One physicist, who was also a Nobel Prize winner, told a reporter of what he perceived was the general scientific view of Apollo: "Not one of my scientific friends thinks it's worth a graduate student's time." Many scientists believed, much like Eisenhower and Kennedy's scientific advisors, that unmanned probes and satellites would be of greater benefit to the scientific community. "We should not forget," said Ralph Lapp, a physicist who worked on the Manhattan Project, "that the single greatest discovery of the space age, the Van Allen radiation belt, was made with only a 30-pound payload." The editor of *Science*, the journal of the American Association for the Advancement of Science, Philip Abelson, told Congress that there was nothing "magical about this decade—the moon has been there a long time and will continue to be there a long time." In an effort to placate objecting professionals, NASA added six scientist-astronauts in 1965, one of whom, geologist Harrison Schmitt, would fly to the moon in 1972 on Apollo 17. Others would be included on flights to the nation's first space station, Skylab, in 1973 and 1974. This helped soothe dissenting scientists to some degree.[75]

Despite the doubts of the so-called experts, surveys found the American people eager to carry on. According to polls, more than 50 percent of Americans thought NASA was spending enough or even too little on Apollo. So there was still considerable support for the lunar program in the public mind in 1966.[76]

■   ■   ■

Apollo would move forward, much to the relief of many astronauts, especially those getting ready for the first Apollo flight. Despite the criticisms floating around, Gus believed in NASA's mission "wholeheartedly." Why? "Let me put it this way," he wrote in his only book, published after his death, "If there had been no Mercury program, there would have been no Gemini program; if there had been no Gemini program, there could be no Apollo program, at least not within a meaningful time. If there were no Apollo program, it could well be that we would be handing on to the next generation a universe dominated by other powers, a world controlled from space, whether we liked it or not." Thinking of the future, he said, "I don't want my two sons to inherit that kind of a world. It is as basic as that."[77]

It had been the hope of NASA to get Apollo off the ground by the end of 1966, and there was even talk of launching Apollo 1 at the same time as Gemini 12 was in orbit and to rendezvous both spacecraft. But the delays and problems with Apollo pushed the launch date back to December 1966, then to January 1967, then to February. Webb tried to keep everyone focused on the realities of the upcoming moon shot. Soon after the completion of Gemini, LBJ hosted Webb and the Gemini 12 crew—Jim Lovell and Buzz Aldrin—at his ranch in the Texas hill country. At a press conference, Webb interposed a word of warning about the start of the moon program. "The months ahead will not be easy as we reach toward the Moon." Reporters thought this might be a sly way of downplaying the significance of Kennedy's deadline, that there might be a possibility of failure. The president, though, explained that it only meant that Apollo would be far more complex than Gemini and that there was a lot of testing ahead. Indeed there would be.[78]

Despite the delays and the problems with the spacecraft, and the souring view of the scientific community, liberal writers, and members of Congress, NASA and their supporters across the country were looking ahead to a very successful 1967, hoping for a greater year than they had enjoyed in 1966. "Looking back, 1966 was NASA's best year," wrote Deke Slayton. "We were in the middle of Gemini, Apollo was building up. Unmanned programs like Ranger, Surveyor, and Lunar Orbiter—all of them designed to support Apollo—were on track. A new program to follow Apollo—Apollo Applications—was being laid out, to perform lunar landings and extended stays on the surface of the moon, in addition to operations with orbital workshops in earth and lunar orbit. The agency would never have as much support as it had in 1966. Or as clear a goal." Things looked very good. Even the Soviets were unusually quiet, without a man in space for nearly two years. Perhaps they had quit the race, and America was about to take a victory lap.[79]

Then came January 27, 1967, and what had always been seen as a routine part of astronaut training and spacecraft evaluation: the plugs-out test. It was "not classified as high-risk," wrote Gene Kranz, a flight director in Houston. Tests were called high-risk and dangerous when fuels or any live explosives were involved. In the plugs-out test, the Saturn booster would not be fueled. In the words of Frank Borman, it shouldn't have been "any more dangerous than taking a bath." What happened that day was unforeseen yet catastrophic, not only for the lives lost and those forever affected, but for the entire nation, for it could have easily galvanized enough opposition forces to end the race to the moon and cancel the entire space program.[80]

IV

THE FIRE:
"WE'RE BURNING UP!"

Lola Morrow noticed something different among the crew of Apollo 1 early on January 27, 1967. Like most astronauts, Gus, Ed, and Roger always seemed to be happy whenever they arrived at their offices at the Cape for training in their spacecraft, still thrilled to have been assigned to the first Apollo flight. The crew had been pushing toward launch day since the previous March when NASA announced their selection. They had faced problem after problem and delay after delay, which was not unusual in such a complex and risky business, but now they were making the final turn and could see the finish line, a launch date of February 21.

But on that winter day in January, Morrow thought that something didn't ring right. As the secretary for the astronaut office, she knew them as well as anyone. "In the morning when the crew came into the office, I sensed something. I don't know what it was that I sensed but I picked up something from all three of them. There was a quietness about them. Instead of being ready for a test where they usually just get up and bounce out the door, it was as if it was something they didn't want to do. Their attitude was one-eighty from anything I've ever seen before." Gus was not his usual chipper self, she noted, while Ed "seemed preoccupied" and Roger "sort of shot by."[1]

Did they have a premonition about what was to come? Gus seemed to have more of one than Ed or Roger, although they too were well aware of the risks. Gus had once said to his pal Deke Slayton, "If there's ever a fatal accident in the program, it's likely to be me." Wally Schirra, who considered Gus a close friend and was the executor of his will, knew that Gus had a different outlook on death than he did. "He thought about death, talked about 'busting his ass,' and had mementos he wanted to pass around in case he died. Perhaps he anticipated disaster and needed to express it."[2]

Just days before the test, when asked about the risks of the first Apollo flight by a CBS reporter, Gus responded, "You sort of have to put that out of your mind. There's always a possibility that you can have a catastrophic failure, of course. This can happen on any flight. It can happen on the last one as well as the first one. You just plan as best you can to take care of all these eventualities, and you get a well-trained crew, and you go fly."[3]

His family was not off the hook either. Gus told Betty that he didn't want everyone mourning him if something bad happened. "If I die, have a party," he said. "If something happens to me, I don't want people sitting over here, crying." She agreed, at least in the moment. "Okay. We'll have a party."[4]

As military pilots, Ed and Roger had a similar outlook. "People might look at our work as being perhaps dangerous, or risky of sorts, but I think we train in it and work in it so much and understand it well enough that we don't look at it from this viewpoint. We accept the risks," Ed said. So did Roger. "I guess when a fellow climbs into a spacecraft, straps himself in and starts waiting for the countdown, he could give what's coming some really serious consideration, but I'm not afraid. I feel a capable pilot should be able to meet the emergencies that may develop. What's more, there's a risk to flying an ordinary plane, just as there is to driving a car, walking across a street or going down a stairway." To the same CBS reporter, Roger said he knew there were "a lot of unknowns and a lot of problems that could develop or might develop and they'll have to solved. And that's what we're there for. This is our business…to find out if this thing will work for us." Roger even told his father not to get emotional if things went horribly wrong. "Dad, if anything happens and I buy the farm, I don't want you to be bitter. I want you to do what you can for the space program."[5]

These attitudes were neither egotistical nor macho talk, but a revelation of how fighter pilots, and astronauts, saw the world. "The rate of progress is proportional to the risk encountered," Neil Armstrong once said, "but to limit the progress in the name of eliminating risk is no virtue." These men were not typical citizens but "a category of men who roll the dice," who "put their hides on the line every day," wrote astronaut Walt Cunningham, who was on the Apollo 1 backup crew. "It wasn't so much that we accepted the risks, but rather that we never admitted the risks existed in the first place." They were a breed apart.[6]

As Gus wrote in his lone book, the men developing systems to fly to the moon were trying to "advance the state of the art," as he called it. Writing in the months before the fire, during the period in which there were problems stacked on top of problems with the spacecraft, Gus noted that there were "still quite a few aspects of space that don't, at the moment, seem to have any hard and fast scientific explanations. Sometimes a rocket will seem to defy all the laws of aerodynamics and simply break up for no apparent reason," he wrote. "Trial and error finally produce a mix that works, but nobody is quite sure *why* it works. Medicine still calls itself an art for the same reasons." And the same could be said for a machine as advanced and as complex as the Apollo command module, and the Saturn boosters that would propel it into space. Any one of them could break down at any moment for any number of reasons, known or unknown.[7]

Most pilots rarely, if ever, talk about failure and death, but they all understand that there are risks involved, especially for test pilots, who know, whether they admit it or not, that every single time they step into a cockpit could be the last time. Yet the attitudes expressed by the crew of Apollo 1 were responses to the possibility of getting killed while *flying*—perhaps during a catastrophic explosion during launch or getting lost in space. No one was thinking about a disastrous failure while sitting on the ground. No one.

■　　■　　■

Facing the Apollo 1 crew on that cold, winter day was one of the final tests before they could be ready for liftoff in a little more than three weeks. It was known as a "plugs-out" test, nothing less than a full-scale launch simulation. The spacecraft sat atop its Saturn 1B booster, which was not fueled since it was not leaving the planet that day. Everything else in the Apollo capsule ran on its own power, not plugged into outside sources, so the plugs were literally out. The astronauts were in their space suits, helmets locked in place. The 100 percent atmosphere inside the sealed capsule was at 16.7 psi, about two pounds higher than the pressure outside the spacecraft, so that leaks could be detected. The crew would go through everything just like a real launch, communicating with launch control, flipping the right switches at the appropriate times, and running the clock down to zero. If the spacecraft passed the test, the mission would be a "go."

The test, though, could be long, boring, and quite monotonous, and that was with a spacecraft in perfect condition. And Spacecraft 012 was far from perfect condition. The day before, January 26, Schirra and the backup crew successfully completed a "plugs-in" countdown test, where the spacecraft was operating on outside power from the ground. "We had performed virtually the same test the night before, with the persistent irritation of 'glitches,' those minor problems we had almost come to expect," wrote Cunningham. "We kept telling ourselves, as did the engineers, that we couldn't expect the first machine out of the factory to be bug-free. Wally had complained that the same problems kept showing up. In the push to keep on schedule, they were not being completely and fully resolved." Although the test wasn't horrible, the prime crew now had to see if the "plugs-out" test would go any more smoothly.[8]

The director of flight crew operations, Deke Slayton, was down at the Cape for the tests. Slayton had talked to Gus about all the problems with the spacecraft, especially regarding communications. "As far as the astronauts were concerned," wrote Slayton, "the big bitch—we had a lot of big bitches on that spacecraft—was that communications were pretty lousy." One idea was to have Slayton get in the capsule with the crew to help with the communications problems. But they quickly determined that it wouldn't work out, so Slayton would remain at the Cape for the test, watching from the blockhouse, which was about 1600 feet away from the pad.[9]

The day before the test, Gordo Cooper had a long talk with Gus and, like Morrow, noticed something a little different in his friend. Gus "wasn't his usual buoyant self," he said. "He thought his mission had a 'pretty damn slim chance' of going its full fourteen days. With three weeks before launch, he was agonizing over the condition of his spacecraft. Some sixty major discrepancies had been identified, and there was no time to fix them all before the 'hot tests' scheduled for the next day at the Cape."[10]

Gus called Betty that night to check on the boys and see if everything was okay at home. He told her that day's plugs-in test was a success and he would fly back home on Saturday morning after his plugs-out run. The astronauts and their wives were invited to a party on Saturday night, January 28, cosponsored by *Life*, which had a contract to publish the astronauts' life stories. "We'll see you on Saturday then," Betty said before hanging up. That was the last time she would ever speak to her husband.[11]

The day of the test, Gus, Ed, and Roger ate breakfast in the crew quarters with Shea, Slayton, and Schirra. "There had been a lot of communications problems with the spacecraft, and Gus wanted to hassle Joe a bit, chew on his ass about them," Slayton said. In his management position in Houston, Shea knew all about the problems. In fact, it was his job to pressure North American to solve them. Six weeks before the fire, at a news conference, Shea admitted that there were problems with

Apollo, more than twenty thousand discrepancies or failures of one type or another at one time or another. "We hope to God there is no safety involved in the things that slip through," Shea had said.[12]

Gus wasn't the only one who felt the spacecraft was unsafe. There were astronauts and other members of the Apollo program who had similar thoughts. Walt Williams, who had been the program director for Mercury, said if the spacecraft was a horse, "they would have shot it sometime in 1966, perhaps as early as 1965." Alan Shepard said it was the "worst spacecraft I've ever seen." Donn Eisele, a member of the backup crew who had actually been slated to be on the prime crew for Apollo 1, called it a "bucket of bolts," which became an all-too familiar refrain. Rocco Petrone used the exact same expression, as did John Young. Training in another Block I spacecraft in Downey on the day of the fire, support crew member Young barked, "Go to the Moon, hell. This bucket of bolts won't reach Earth orbit." A North American Aviation quality control inspector, Thomas Baron, characterized Spacecraft 012 as "sloppy and unsafe." Baron's name would come up later in the fire investigation.[13]

Schirra, as the backup commander, had an uneasy feeling as well. The day before the test, after the backup crew had completed the "plugs-in" test, he took Gus aside and told him that the spacecraft "just didn't feel right," and if he had "the same feeling I do," then he should get out of the capsule. "Listen to me, Gus. It'll take you a minimum of ninety seconds to get all those hatches open. If you have a problem, even a communications problem, get out of the cabin until the problem is cleared. Got it?"

"Got it," Gus replied. But after having lost *Liberty Bell 7*, there was almost nothing that could have caused Gus to bail out of the Apollo spacecraft prematurely that day or any other.[14]

■   ■   ■

With warnings in hand, the crew climbed inside their Apollo spacecraft at 1:19 p.m. Schirra and the backup crew flew back to Houston, as did Joe Shea, while Slayton remained to monitor the test from the block-house, but he did drive out to the pad with the crew, ride up the gantry to the white room—the enclosed area at the top of the rocket where astronauts climb in and out of the capsule—and remained until they were strapped in, something he had never done before. There were more than one thousand people at the Cape to support different aspects of the plugs-out test, in addition to controllers in Houston who would also be listening in. As was the normal routine, Gus climbed in first and slid over to the commander's seat on the left side of the spacecraft. Roger crawled in second and was strapped into his seat on the right. Ed, as senior pilot, occupied the center seat so he got in last.[15]

Then came the glitches. As the crew of Apollo 11 later wrote, "it was one of those days when nothing seemed to be going right." During the test, Lola Morrow listened to everything on the squawk box, and she could tell things were not right, that everything seemed to be going wrong. "There were too many problems. I kept thinking to myself— 'Project Apollo. We should call it Project Appalling.'"[16]

As the test began, there were yet more problems with the environmental control unit, the same unit that had been replaced twice in recent months. Once inside the spacecraft, and hooked into the oxygen system, Gus started the oxygen flow and immediately complained of an odor of "sour milk" or "buttermilk." Controllers delayed the countdown so that the air could be tested but no discernible problems were discovered. The test resumed, and the hatches were closed and sealed shut at around a quarter to three in the afternoon.[17]

The communications system was also rife with issues that continued for what seemed like an eternity. Static, garbled words, long pauses, and dropouts plagued the lines between the spacecraft and launch control, as were holds in the countdown. "The overall communications problem was so bad at times that we could not even understand what the crew was saying," test conductor Skip Chauvin later told the investigation team that investigated the fire. Near the five-hour mark in the test, one engineer mentioned to Chauvin that it might be a good idea to stop the test. "Let's cancel out today. This could go on forever. We're better off if we shut down and do the full test again." But Chauvin declined. The test would resume.[18]

A communications issue held the count at 5:40 p.m. After a significant delay to work the problem, test conductors were ready to continue the countdown at 6:30 p.m. Gus was growing more and more frustrated the longer the test dragged on.

"How are we gonna get to the Moon if we can't talk between three buildings," he said, his displeasure ringing strongly in his voice.

"They can't hear a thing you're saying," Ed said, his voice showing a slight touch of humor about it all, as if it was no big deal.

"Jesus Christ," Gus moaned.

"I said how are we gonna get to the Moon if we can't talk between two or three buildings?" Gus repeated, sounding even more frustrated.

While this exasperating exchange occurred, somewhere under Gus's seat a chafed wire was about to spark and cause an inferno.

After a minute of staticky silence, just 4.7 seconds after 6:31 p.m., a nd more than five hours into the test, the wire sparked and ignited a fire, a blaze that quickly found ample fuel in the 100 percent oxygen atmosphere and more than enough flammable material. Within seconds, the blaze was spreading like a brushfire.

Sounds of movement in the capsule went over the commlink. Controllers monitoring the test noticed a power surge in one of the electrical buses that power the spacecraft—AC Bus 2—while doctors monitoring the astronauts' biomedical data saw a jump in heart and pulse rates, particularly in Ed.

Slayton, in the blockhouse with Rocco Petrone, saw in their monitors and instruments "messengers from hell." One gauge showed a huge amount of oxygen flowing into the astronauts' spacesuits. Electrical currents went wild, surging madly, while the gauges showing cabin temperatures pushed as far as the needle would go.[19]

Controllers in Houston were seeing the same abnormal indications. Gary W. Johnson, an electrical engineer, was watching his console in the

Staff Support Room at Mission Control. Engineers in the SSR offer technical support to flight controllers during the mission, or in this case for the test. At the time the fire broke out, 5:31 p.m. Houston time, Johnson saw an electrical spike on his console. What could be going on?[20]

They didn't know much more in the blockhouse. "All we could do is listen," Slayton said. "We had a TV camera trained on the hatch window, which normally showed little more than a dark circle. That circle lit up and went almost white."[21]

Then the static was broken by a voice, sounding like Roger, shouting either "Hey!" or "Flames!"

Seconds after Roger was Ed, "Hey, we've got a fire in the cockpit!" he said, as he began the procedures to unlatch the hatch, which was directly over his head, so the crew could get out of the burning spacecraft.

Agonizing seconds of silence punctuated only by static followed as the crew began their emergency procedures. But the fire was growing in intensity with each passing second. Finally, a voice pierced the static. It was Roger.

"We have a bad fire! We're burning up!"

Then there was a final blood-curdling, agonizing scream of pain before dead silence. At this point the crew was likely deceased, not so much burned from the fire itself but from the toxic fumes. The fire had burned though the hoses that fed their spacesuits, so instead of life-giving oxygen, they could only breathe in life-destroying toxic smoke. They perished in less than twenty seconds.[22]

Donald Babbitt, the pad leader who was in the white room with four other technicians, immediately sprang into action when he heard the crew's cries of "Fire!" He yelled to those near the spacecraft, "Get them out of there!" He then hit the alarm button to alert everyone on the gantry of the danger.[23]

Chuck Gay, the test director in the launch control room, began calmly calling to the crew to get out of the spacecraft. "Crew, egress," he said. "Crew, can you egress at this time?" There was no response.[24]

Meanwhile, the pressure was rising inside the spacecraft fueled by the intense heat, estimated at an un-survivable 1200 degrees Fahrenheit. The heat and pressure climbed so high that the spacecraft actually burst, rupturing the hull of the capsule with such force that the launch pad crew in the white room, including Babbitt, were knocked backward off their feet by the blast. Technicians grabbed gas masks and fire extinguishers to try to put out the fire but were momentarily stopped by the powerful flames and thick smoke that made visibility impossible. "You couldn't see six inches from your face," said one.[25]

When the first call came about a fire in the spacecraft, Slayton, Petrone, and Roosa could do nothing. Watching from the blockhouse, thick black smoke billowed from the top of the launch tower, emanating from fire spilling out of the spacecraft rupture. In the white room "it was just mass confusion up there."[26]

But now a new worry quickly got the attention of those on the launch tower, as well as those in the blockhouse and launch control. Although the Saturn rocket was not fueled and wasn't a danger, the escape rocket on the nose of the spacecraft, meant to pull the capsule away from the rocket if there was an explosion, was fueled with a solid propellant. There was immediate concern that the fire boiling out from the ruptured hull could ignite it. And it was no small rocket. By itself, it had twice as much power as the Mercury Redstone that Gus had taken on his suborbital mission into space in 1961. If it ignited, said Stephen Clemmons, a pad technician, "all of us were cooked." Someone shouted through the mass chaos and smoke, "She's going to blow!" With this threat, the pad crew had no choice but to quickly evacuate the white room.[27]

Technicians from North American Aviation were watching the grue-some events unfold from the white room. One was Bruce W. Davis. "I heard someone say, 'There is a fire in the cockpit.' I turned around and after about one second I saw flames within the two open access panels in the command module near the umbilical." A spacecraft-systems engi-neer from North American, Jessie L. Owens, was standing near the pad leader's desk. Hearing someone shout "Fire," Owens thought he heard what might have been the cabin relief value opening and gas escaping at a high rate of speed. "Immediately this gas burst into flames somewhat like lighting an acetylene torch. I turned to go to the white room... but was met by a flame wall."[28]

Seeing the catastrophe unfold right in front of him, Babbitt grabbed a man with a headset and screamed into it, "We're on fire! I need firemen, ambulances, equipment, now!" Many of the other personnel on the launch tower sprinted to level 8, where the gantry and white room were located. They grabbed masks and fire extinguishers and took on the fire, fighting it back until they reached the hatch, largely due to their heroic efforts but also due to the fact that the fire began to burn itself out. The hatch handles were so hot that they burned through the gloves of the technicians who were wrestling to open the cabin. They continued hosing it down to cool it enough so it could be opened to see if the crew had somehow survived. But it took nearly five and a half minutes for them to open the hatch. By then, Slayton wrote, "it was obviously too late." Pad personnel had done everything they could to try to save them but ultimately could not. Twenty-seven men were treated for smoke inhalation, while two had to be hospitalized.[29]

A North American Aviation representative, Jim Pierce, was also on-hand to watch the test. When he saw the disaster unfold, he immediately dialed the headquarters in Downey and connected with "Stormy" Storms. On speakerphone in the company conference room, Pierce described the horror he was witnessing. "There's fire spewing from the spacecraft... I can see molten metal falling away! The whole thing could blow up any minute!" Hearing the news, someone listening in responded, "Oh, Jesus!"[30]

When Schirra and the backup crew arrived back in Houston and walked into Mission Control, they found "shock, confusion and consternation." NASA's lead flight director, Chris Kraft, listening to everything on his headset, had heard the agonizing cries coming from inside the spacecraft. "I heard their screaming voices in the cockpit of the spacecraft," Kraft recalled. "I heard them scream that they were on fire.

I heard them scream 'get me out of here.' And then there was dead silence on the pad."[31]

Back in the Staff Support Room (SRR), Gary Johnson waited for word of what was happening at the Cape. "Pretty soon Chris Kraft came running into the SSR and said, 'We need to play back our data so everyone can review it.' And then he told us no phone calls out of the building. We knew something crazy was going on out at the Cape, but I kept thinking since the crew were in their spacesuits, they should be okay. I was holding out hope."[32]

But hope didn't last long. "Within minutes we knew they were dead, and we were in deep, serious trouble," Kraft said. "Nobody really said anything for 15 minutes, until they got the hatch open. We were sitting there, waiting for them to say what we knew they were going to say."[33]

Joe Shea was back at the MSC as well. He walked into his office on the seventh floor at 5:30 p.m., which was 6:30 p.m. at the Cape. "It was about twenty minutes before we realized that they were not about to get out," he said. "It was pretty tough."[34]

Slayton had ordered Fred Kelly, the flight surgeon, to the white room when the fire broke out. Kelly, though, knew what would await him. Slayton raced to the pad too. Once the fire was out and the hatch was cooled enough to open, Pad Leader Babbitt shined his flashlight into the burned-out spacecraft and found the lifeless bodies. They were in exactly the position they should have been. Roger was sitting in his seat, while Gus was slumped over on top of Ed, who was trying to get the hatch open. These positions indicated that they relied on their training, even as flames raged around them, and worked the problem by the book. Roger, in the number three seat, was to remain in his position in case of emergency and maintain communications with the ground, which is why his lone voice cried out at the end of the recordings, while Ed, assisted by Gus, worked to get the hatch open. They didn't panic, as regular people likely would have, but showed their true mettle as military-trained pilots, even as they burned alive.[35]

A horrified Rocco Petrone, sitting in the blockhouse watching the television monitor, got on the headset to talk to someone in the white room. "Can you do something for the guys?" he asked. The answer he got was one he didn't want but knew he'd get. "No...no...it's too late." Doctors were rushed to the spacecraft, as routine procedure, but also for the faintest glimmer of hope that perhaps somehow, in their suits, one or more had survived. When they arrived and peered into the capsule, everyone knew. One looked at Slayton and shook his head. "They're gone," he said. When Slayton looked inside the spacecraft, he said it looked "like the inside of a furnace." There was no chance they could have survived.[36]

■   ■   ■

No one will ever know if they could have lived with an easier means of escape. But they did not have one. The complex hatch did them in every bit as much as the flames, fumes, and smoke. In one of the great ironies in space program history, if the hatch had been equipped with explosive bolts, like the hatch on Gus's Mercury capsule, they could have been out of the burning inferno within seconds. But the hatch concept had been changed, largely because of the incident with *Liberty Bell 7* five and half years before. They couldn't risk a malfunction like that in the vacuum of space, so the explosive bolts were removed from the hatch design.

The hatch would become a major focus during the investigation. It was not a simple, one-piece mechanism that could be opened quickly, but was actually three separate hatches—a heavy inner hatch that opened to the inside of the spacecraft, an outer hatch that opened to the outside, and an outer cover, as part of the tower jet safety system, that covered over the entire capsule. To get out, Ed, occupying the center seat, would insert a ratchet device and loosen the hatch in six different places, then lower the extremely heavy hatch down to the floor of the spacecraft, then open the outside hatch. Someone from the ground crew would have to open the outer cover. It was a very cumbersome process that took several minutes to execute in normal circumstances. A raging fire was hardly normal. And, to make matters worse, with far greater pressure inside the capsule pressing against the hatch, it was all the harder to get open. Even as strong and athletic as Ed White was, he could not do it in time.[37]

Soon after the tragedy struck, NASA put out a very brief, and very vague, press release. It read: "An accidental fire has broken out on the Apollo launch pad, killing at least one person. The space agency says its victim may have been one of the three astronauts scheduled to make the trip." Although such a deliberately evasive, and misleading, statement may seem like a rather cruel way to respond, it was done to inform the public of the disaster before news agencies got ahold of the story. NASA wanted to make sure that the wives and families were informed of the death of their loved ones by team members, not the television or radio. Jack King, the chief public affairs officer for NASA, "didn't want to spend the rest of [his] life knowing that Betty Grissom heard about [the fire and her husband's death] because of [his] announcement." Once the families were informed, the full story was released to the public.[38]

Slayton left the pad and went back to the crew quarters to start making the most important phone calls he would ever make in his life, with the help of Morrow. "Every telephone in the office was ringing," Morrow said. "The whole world was calling." She recalled that when Slayton arrived he was shaking so much he couldn't light his cigar. He had the toughest of tasks ahead of him, to inform all three wives, who were now widows. Slayton called the astronaut office in Houston where he got Schirra on the phone. He'd heard about the fire already after he and his crew landed at Ellington Air Force Base near Houston and was informed of the disaster. So he had already started the process of rounding up some astronauts and their wives to go be with the Betty, Pat, and Martha.[39]

Slayton also had to call out to North American in California to alert the flight crews who were conducting tests of their own. He spoke with astronaut Al Worden, who was in the astronaut office at the Downey plant that day. The crew of Tom Stafford, John Young, and Gene Cernan, who were part of the support crew for Apollo 1 and the backup crew for the second manned mission, were, at that moment, training in an identical Block I spacecraft at North American, and fussing over the same problems. After the fire and subsequent investigation, they would understand more clearly that they "were a ticking time bomb inside that spacecraft, one that could go off at any time." On the phone with Worden, Slayton was "terse and business-like." He told Worden about the tragedy and ordered him to get Stafford's crew out of the spacecraft immediately. All tests were cancelled. The four were to return at once to Houston.[40]

Slayton also needed to contact NASA officials. Interestingly, on that very day, NASA administrator Jim Webb and other leaders were at the White House for the signing of the Outer Space Treaty, which established international laws to govern space. The exploration of outer space would be for peaceful purposes and not for military applications. In total, fifty-seven nations were represented, including the ambassadors of Great Britain and the Soviet Union. Several astronauts were also in attendance, including Neil Armstrong, Jim Lovell, and Gordo Cooper. "This is an inspiring moment in the history of the human race," the president said.[41]

That evening, after the signing ceremony, Webb met with the Apollo Executives Group at the International Club, which was close to the White House. The meeting consisted of the upper management of NASA—Webb, von Braun, Gilruth, George Mueller, and Kurt Debus, the director of the Kennedy Space Center at the Cape—and the chiefs of several major aeronautical companies: North American Aviation, McDonnell, and Grumman. Vice President Humphrey was also in attendance. The subject of the meeting was the lessons that had been learned in Gemini and what problems the program faced with Apollo. General Samuel Phillips, head of the project in Washington, had planned to deliver a report to the group entitled "Apollo Program: Top Ten Problems."[42]

Around 7:30 p.m., as the group was enjoying cocktails, urgent messages began coming in. Debus took Webb aside and told him the news about the fire. Webb immediately phoned the White House and spoke to one of the president's secretaries, Jim Jones. He dictated a short statement that was to be handed to the president, who was upstairs in the residence attending a party for his retiring commerce secretary, John Connor. Handed the folded note, Johnson was devastated. Lady Bird Johnson wrote that night in her diary: "I watched his expression as he read it. His face sagged and my heart lurched. I knew the news was something bad and something close." The president then read the statement aloud to the gathering. "I have a sad announcement to make," he said. "James Webb just reported that the first Apollo crew was under test at Cape Kennedy and a fire broke out in their capsule and all three were killed. He does not know whether it was the primary crew or backup crew but believes it was the primary crew of Grissom, White, and Chaffee." The note struck Johnson hard. "The shock hit me like a physical blow," he said, and "the happy atmosphere changed to one of stunned grief." He excused himself from the party and went to the Oval Office. "All through the night I thought in anguish of those brave young men and of their bereaved families."[43]

General Phillips got on the phone and called Kennedy Space Center. Knowing an investigation would be coming, he wanted everything seized and held. "Impound all records, all tapes, every last piece of equipment associated with the fire, and I mean everything," he ordered. "Put the clamps down everywhere. I'm on my way."[44]

But word was spreading quickly. Robert Seamans, NASA's deputy administrator, did not attend the Apollo Executive Group meeting in Washington but had a previous engagement at his home. When he arrived around 7:00 p.m., his phone was already ringing. Hearing about the tragedy, he went to his office at NASA headquarters and began making calls to find out all he could about the fire. While on the phone with the secretary of defense, Bob McNamara, an operator broke in with an emergency call. It was Peter Hackes of NBC News. "This is a national emergency," he told Seamans. "The word is out. The country is almost in a panic, and you've got to go on TV and reassure the public!" But Seamans refused. He needed more information.[45]

Jim Lovell had been at the White House event. When he got back to his hotel room, his red message light was blinking: he was to call Houston immediately. When he phoned the center, he got someone on the phone that he did not know, perhaps a public relations guy who was tasked with briefing the astronauts.

"The details are sketchy but there was a fire on pad 34 tonight," he told Lovell. "A bad fire. It is probable the crew did not survive."

"What do you mean 'probable'?" Lovell asked. "Did they survive or didn't they survive?"

"It is probable the crew did not survive," was the tepid response.

Lovell was stunned and asked who knew about the tragedy at this point. Only those who needed to know were informed, he was told. In a word of caution, Lovell was ordered to stay put in his hotel room at the Georgetown Inn, and not to venture out, as the media would be on the hunt for anyone from NASA to get comments about the fire, especially astronauts.[46]

Like Lovell, Gordo Cooper, another Mercury astronaut and a close friend of Gus, had been at the White House event and also had retired early to his room at the same hotel when he received a phone call from Congressman Gerald Ford, a friend and member of the House Space Committee.[47]

"Gordon, I just got word that there's been an accident at the Cape and the crew was killed," Ford said.

"I felt my mouth go dry," Cooper recalled. *The hot tests...the sixty major discrepancies*, he thought, his mind immediately going back to his conversation with Gus the day before.

He then managed to ask Ford if he knew any details. "Pilots always want to know the details of an accident," Cooper later wrote. "It is a hard habit to break, even as I struggled with the emotions of knowing I had lost my best friend." Ford didn't know any.

Lovell eventually left his room, his mind in a thousand directions, and ran into his astronaut colleagues who had also attended the treaty-signing event at the White House and were staying at the Inn. They too had heard about the fire.

"What the hell happened?" Lovell asked Armstrong, Cooper, and Dick Gordon, who had flown on Gemini 11.

"What happened?" Gordon replied. "That spacecraft happened, that's what. They should've deep-sixed that thing long ago."[48]

America's first man in space, Alan Shepard, was in Dallas, Texas, at a dinner where he was about to give a speech. Just before taking the podium, someone whispered in his ear that there had been a fire on the pad and Gus, Ed, and Roger were dead. It hit him like "the force of a sledgehammer." He stammered to the podium in what can only be described as a fog. "I...I have just been informed of the loss...the loss of my comrades," he said, fighting back tears.[49]

Rookie astronaut Charlie Duke, who had yet to fly but would walk on the moon on Apollo 16, was back home in South Carolina, receiving a Jaycees Young Man of the Year Award. "I was actually in the middle of my acceptance speech when I was called away from the microphone to be informed of the terrible news. It's hard to describe the shock and sense of loss I experienced at this point. It was almost with tears that I walked back into the meeting, choked up and hardly able to talk, and announced the death of these three friends." The future of Apollo looked bleak to Duke. "That evening with real sadness I told everybody, 'Now it seems there is no way we are going to make it to the moon by 1970.'"[50]

Flight Director Gene Kranz, who would become legendary for his heroic efforts with Apollo 13, had been monitoring the early part of the test at Mission Control but left after his shift was over so he could go to dinner with his wife Marta. There was a knock at his front door. Chris Kraft, running things at the MCC, had shut down all outgoing calls, so Kranz had no idea what was going on. He opened the door to find a flight controller, Jim Hannigan, who walked in "agitated and breathless," Kranz said. "They had a fire on the launch pad," Hannigan told him. "They think the crew is dead!"[51]

"I had a sudden apocalyptic vision of a gigantic explosion that had taken out the flight crew, the Saturn rocket, and the launch complex," Kranz wrote later. He raced back to Mission Control, where Kraft told him what he had heard from Slayton. The vision Kranz had could easily have come true. "Deke thinks we were damned lucky that we didn't lose a hell of a lot more. There was fire coming from the capsule, molten metal dribbling down the side of the service module," Kraft said. He'd been around since the beginning and was as upset about the tragedy as anyone. "I'd dealt with flight-test deaths before, but this was different," Kraft wrote. "Now we'd put three astronauts into harm's way and made their escape impossible. They were dead and we knew it was our fault. The fire on the pad and its consequences would be profound." And profound could mean the death knell for the program.[52]

Careful not to disturb the evidence in the spacecraft, medical personnel did not remove the deceased crew until after 1:00 in the morning. The bodies were sent to the Biomedical Operational Support Unit about a mile from the pad. An Air Force medical clinic, it would act as a temporary morgue. Seamans reviewed photographs of the bodies, both inside the spacecraft and once they were at the morgue. The "experience wasn't pleasant," he said, "but it wasn't as gruesome as might be imagined either." The fire had destroyed 70 percent of Gus's spacesuit, 25 percent of Ed's, and 15 percent of Roger's, indicating that the fire had been worse on Gus's side of the capsule. Later that morning, January 28, an autopsy revealed that they did not burn to death but asphyxiated due to the carbon monoxide that surged through their breathing system and right into their lungs. The burns they sustained were likely survivable, and probably postmortem.[53]

■　■　■

Soon after the fire, astronaut wives began making the difficult sojourn to be with their closest friends, Betty, Pat, and Martha. Living close together, with husbands in the same dangerous occupation, astronaut wives had a special, close-knit relationship. In fact, it even had a name: "Togethersville." No matter what, they were in it together. They supported each other through thick and thin, and most especially in tragic times such as these. The astronauts themselves would also be supportive but nothing could match the closeness felt by the wives.[54]

Betty Grissom was at home that Friday evening, enjoying company and an adult beverage with Adelin Hammack, whose husband worked at Mission Control, when there was a knock at the door. It was Jo Schirra. "There's been an accident at the Cape," she said. "I think Gus was hurt."[55]

But Betty had the feeling it was much worse. "It's over," she thought to herself. After a few minutes, the astronaut physician, Dr. Chuck Berry, came to give her the bad news. Gus was dead. Betty, though, didn't break down and fall to pieces. And it was not because of what Gus had told her. She had prepared herself for this moment, unlike most other astronaut wives who somehow believed their husbands couldn't get killed in such a risky business. "I faced Gus's death 100,000 times," she said. "When he flew in combat in Korea, when he was a flying instructor, when he made his Mercury flight, and the Gemini flight...all the flights. Gus never tried to keep it from me, to 'shield' me. I knew what it was." In her book, *Starfall*, published seven years later, Betty wrote, "I didn't feel fatalistic" about the possibility of Gus's death, understanding "that sooner or later it was bound to happen. I was never really anxiously concerned. I thought of it, but only now and then. I don't think you could have survived if you had dwelled on it."[56]

Others noticed what can only be described as a certain toughness she had. Betty was "stone-faced and composed," said Rene Carpenter, wife of Mercury astronaut Scott Carpenter. "Betty Grissom was the only one who didn't cry," wrote Walt Cunningham. She was "stoic, composed," like a "veteran. She had traveled a lot of miles with a much rougher guy than had the other wives." She "kept her calm and dignity" throughout.[57]

Pat White had taken daughter Bonnie to piano lessons and arrived home later than usual, but when she pulled into her driveway she saw Jan Armstrong, her neighbor and close friend, waiting for her. Soon Pete Conrad and his wife arrived. Unlike Betty, Pat's grief "was visible," Cunningham said, "She took it hard." Of the three, she suffered "the most extreme case of the pad fire's aftereffects," wrote Frank Borman, who came by later with his wife Susan, who was close to Pat and would stay with her for hours while she cried. "Who am I, Susan?" she asked. "Who am I? I've lost everything. It's all gone." Sadly, in 1982, after a terminal cancer diagnosis, Pat White took her own life. "It's fair to say Mom really never got over his loss," wrote Ed III and Bonnie Lynn. For Borman, "She was the last victim of Pad 34."[58]

Martha Chaffee was busy clearing away the dishes from feeding the kids their supper when her doorbell rang. It was Sue Bean, wife of astronaut Al Bean. "I thought you might like some company," she said. In a few moments Clare Schweickart came by. They told her there had been an accident but they did not know any details yet. Martha then phoned Roger's room at the Holiday Inn. It couldn't have been him, she said, because "he isn't flying tonight." But she got no answer. Then Mike Collins arrived. And Martha knew. "Mike, I think I know, but I have to hear it." Cunningham and his wife came by to see Martha, and she "was still glassy-eyed and had obviously not yet come to believe that Roger had really died." But Martha, like Pat and Betty, still had the toughest of jobs ahead—telling their children that their fathers would not be coming home again.[59]

President Johnson sent personal letters to the families of all three fallen astronauts and issued an official statement from the White House on the evening of the fire. "Three valiant young men have given their lives in the nation's service. We mourn this great loss and our hearts go out to their families." Vice President Humphrey, who held LBJ's old position as chairman of the Space Council, and who often visited the Cape and knew many of the astronauts quite well, issued his own statement. "The deaths of these three brilliant young men, true pioneers and wonderfully brave, is a profound and personal loss to me. I have had such close relationships with them that my sorrow is very deep. My heart goes out to their families and loved ones." Yet Humphrey was determined that the program would go on. "The United States will push ever forward in space and the memory of these men will be an inspiration to all future space-farers."[60]

Jim Webb echoed the vice president. "We in NASA know that their greatest desire was that this nation press forward with manned space flight exploration, despite the outcome of any one flight. With renewed dedication and purpose, we intend to do just that. I have extended my sympathy and that of all employees of NASA to the families of Astronauts Grissom, White, and Chaffee," he wrote. "The nation tonight feels a great sense of loss. That feeling is even greater among those of us who worked with those competitive young men who were so completely devoted to enlarging man's capability in space flight."[61]

Secretary of Defense Robert McNamara, the administrator in charge of the nation's military forces, expressed the sentiments of all service members. "Our brave men in uniform, whether in Vietnam or seeking the frontiers of the future, mourn with all of us the tragic loss of three gallant and dedicated American airmen. To the families of Lieutenant Colonel Grissom, Lieutenant Colonel White, and Lieutenant Commander Chaffee we send our deepest condolence."[62]

Former President Dwight Eisenhower, though a foe of the moon program, issued a statement of sympathy. "The accident that took the lives of three of our highly trained, skilled and courageous American astronauts is a tragic loss to our entire nation. Mrs. Eisenhower and I extend to their families our deepest sympathy. Our thoughts and prayers are with them."[63]

Even the famous were deeply affected by the loss. On January 17, ten days before the fire, Gus, Ed, and Roger were leaving the North American plant in Downey, California to fly back to the Cape when one of the T-38 jets they were assigned had a malfunction that required maintenance, so they stopped at Nellis Air Force Base near Las Vegas. Awaiting the repairs, the trio, donning their flight suits and jackets, decided to take in a show featuring Frank Sinatra, a fan of the space program. "No sooner had Gus, Ed, and Roger appeared in the room than Frank had them brought up to a front table," Al Shepard recalled. Sinatra introduced them and took a liking to Gus's jacket, complete with mission patches and logos. "Here, take it," Gus said, pulling off his jacket and handing it to Sinatra, who was "so moved he cried before his audience. Ten days later, he cried a lot more."[64]

The media, of course, were quick to jump on the tragedy, as the news circled the globe and turned into a major story. Walter Cronkite said on CBS News, "This is a time for great sadness—national sadness and certainly the personal sadness of the people in the space program. But it's also a time for courage. And if that sounds trite, I'll change the word to guts.... These guys who went into it knew it was a test program...[that] was bound to claim its victims.... It should not be a cause for our turning back or having any question of faltering in our progress forward toward the landing on the Moon.... It shouldn't in any way damage our national resolve to press on with the program for which these men gave their lives."[65]

But much of the coverage was very negative. The famous columnist Walter Lippmann, writing for *Newsweek*, decried the "competitive timetable" of the "artificially accelerated program.... At the risk of their lives, these men are being sent on a mission for which the scientific preparation is far from adequate." The *New York Times* likewise chastised the "technically senseless" and "highly dangerous" decision to put a man on the moon before the end of the decade. "How could those in charge of the test have 'failed to identify it as being hazardous?' The three astronauts had been put into what even a high school chemistry student would know was a potential oxygen incendiary bomb, one needing only a good spark to initiate catastrophe," the editors wrote, "Even hospital workers know the dangers of oxygen fires when they put a patient on it." The *Washington Sunday Star* criticized the accelerating costs of the program and was one of the first to condemn the choice of North American Aviation as the prime contractor to build the spacecraft, opining that "know-who" seemed to win out over "know-how."[66]

The Soviets issued statements of sympathy and paid tribute to Gus, Ed, and Roger, yet they also expressed their opinion that the accident was the result of "haste" on the part of America and "flaws" in the design of the Apollo spacecraft. *The Nation* magazine rebuked NASA for its "gotta-beat-the-Russians" philosophy and agreed with the Soviets on the cause of the disaster. "The charges have substance, and independent-minded aerospace commentators know it. As recently as late December, 1966, the NASA 'success schedule' called for a manned lunar landing in late 1968 or early 1969. These target dates were set to outpace the presumed Soviet timetable and to live up to President Kennedy's rather foolhardy deadline of 1970," the editors opined. Although *The Nation* pointed out its support for Apollo, there was an element of opposition: "What we do oppose is the conversion of a great undertaking into a juvenile contest and a test of American ascendancy in the cold war. Whether U.S. or Soviet astronauts are the victims, this competition is senseless from any humane, scientific or even political standpoint."[67]

But astronauts and test pilots knew better. "To the NASA of those epochal years, 'success' and 'safety' were interchangeable words," wrote Frank Borman. "Yet NASA also recognized the existence of two other interchangeable words: 'mission' and 'risk.' To maximize safety merely meant reducing risk, without forgetting that some risk—even terrible risk—was unavoidable." The fire, wrote Walt Cunningham, "reminded the American public that men could and would die exploring the heavens." As *Aviation Week* likewise pointed out, tragedy "is an inevitable element of all man's efforts to extend his horizons," and the "exploration of space is no exception." Apollo 1, the editors rightly predicted, "was the first tragedy in the nine-year history of manned space flight, but it will not be the last."[68]

NASA also understood that important, yet uncomfortable fact, and still wanted to push ahead with the program, just as Gus, Ed, and Roger would have wanted. Roger's father, Don Chaffee, himself a pilot, certainly understood those same risks, even though the tragedy was tough to take. "We were conditioned for things like this, but it's still an awful shock," he told a reporter.[69]

The American people, though, were having misgivings, probably lulled into complacency by a "false sense of security," given the "outstanding success and perfect safety record of Mercury and Gemini," *Aviation Week* noted. A Harris poll in April, three months after the fire, showed support for Apollo waning. Forty-two percent of Americans wanted to cut the space program, while only 13 percent thought it should be expanded. Thirty-eight percent said it should remain at its current level. In July, another Harris poll revealed that a majority, 54 percent, did not think the moon program was worth the enormous cost, in both lives and money. Thirty-four percent did. The American people "realized the mortal price tag traveling in space could carry—and they didn't much like it. America might be willing to continue funding NASA's increasingly risky cosmic expeditions, but give them too many flag-covered coffins or too many crepe-draped widows, and they just might drop the hammer on the whole operation," wrote Jim Lovell in *Lost Moon*.[70]

Talking to regular Americans in all walks of life, newspapers sought to get a sense of what most people were thinking. "The program's not worth a man's life," said one. "It seems to me we should stop at this point and reexamine the whole program," said another. But some understood the risks. "In any experimental program like this you're bound to have tragedy," said a police officer in Pittsburgh. "The number of fatalities is infinitesimal, though, when you compare it to something like the carnage on the nation's highways." Given these various opinions, no one could really say what the public mood actually was on this matter. Only time would tell.[71]

The funerals were scheduled for January 31, just four days after the fire, which gave Betty, Pat, and Martha little time to plan. Betty, though, handled the funerals in roughly the same manner as she handled the news of Gus's death. Schirra, as a close friend of Gus and executor of his will, discussed funeral plans with Betty. "What do you want?" he asked. "The whole nine yards," she said. "The whole thing, whatever they do, do it." The funerals for Gus and Roger would be almost identical, only a few hours apart, at Arlington National Cemetery.[72]

Pat, though, was having a more difficult time of it. When the Bormans arrived the night of the fire, Frank Borman found that Pat, dealing with the tragic news that her husband had just been killed in an awful fire inside his spacecraft and having to comfort her children, was also fighting NASA, which had introduced an element of insensitivity to the family's mourning. Washington politicians had already decided that all three fallen astronauts would be buried at Arlington.

But Ed's wishes had always been to be buried at West Point, where he and his father had graduated. "Pat was staging a tearful but losing battle," Borman wrote.

"They told me there has to be only one ceremony," she managed to get out through her tears. Seeing Pat in such a state set Borman off. "I couldn't believe it," he said. "They were worrying about what would make it easier on them than on the victims' families. It was a typical bureaucratic reaction and I was angry."

"That's nonsense," Borman told her. "We're going to do exactly what you want and I'll take care of it." And he did. Borman made phone calls and laid it out succinctly to a bureaucrat high on the totem pole. "Ed White's funeral will be at West Point like the family wants, so you might as well go ahead and arrange things—it's the way it's going to be." And it was.[73]

Gus and Roger were buried with full military honors during separate services at Arlington National Cemetery, attended by President Johnson. Gus's pallbearers were the six remaining members of the Mercury class of astronauts—Alan Shepard, John Glenn, Deke Slayton, Scott Carpenter, Wally Schirra, and Gordo Cooper. For Roger, it would be members of the third group of astronauts, including his good friend Gene Cernan. Johnson spoke to the family members, offering personal condolences. But his efforts were met not with kind acknowledgements but brief glances and slightly nodding heads, as most family members quickly looked away from the man who had done so much to create the space program and point it toward the moon. It wasn't a snub, as some in the media had portrayed it, but the fact that the families were "still in a state of shock," said Gus's brother Lowell.[74]

The first lady and Vice President Humphrey traveled to West Point, New York, to be at Ed's memorial service, which, sadder still, fell on what would have been Ed and Pat's fourteenth wedding anniversary. Pallbearers were Ed's closest friends—Frank Borman, Jim Lovell, Neil Armstrong, Pete Conrad, Tom Stafford, and, of course, Buzz Aldrin. "I can't think of a better symbol of courage for future generations of cadets," he wrote of Ed's burial at West Point. Lady Bird Johnson wrote about the day in her diary. "There was an element of strength and beauty in this cruel day. Mrs. White wept softly as they presented her the folded flag from Ed White's coffin." The first lady then leaned down to speak to Pat and the children and received a better reception than had her husband. Pat then asked the first lady to pass a message to President Johnson. "Please tell the president that Ed loved him. Now will you remember to tell him that?" For the first lady of the United States, the kind sentiment on such a solemn day was "almost too much."[75]

Soon after the funerals, the three widows met with Slayton. They had a gift for him. After an astronaut's first flight, he was presented with gold astronaut wings. Slayton had never flown in space since he was grounded after his heart issue during the Mercury program so he did not have a set of astronaut wings. But Gus, Ed, and Roger had had a set made especially for Slayton that included a diamond. They had intended to give it to him after Apollo 1's successful flight. So the three ladies asked to see Slayton and gave him his wings. "I was still badly shaken. Rattled and battered is a good way to say how I felt. Betty, Pat, and Martha were holding up better than I was," Slayton later said. "They broke the tension by making me a surprise presentation. Since what those guys planned could never happen now, the wives, for whatever reason, chose this, the saddest and grimmest occasion in their lives, to present that pin to me. I was absolutely overwhelmed. Flattened. It was a gesture I'll never forget." He wore them with pride every day, with the exception of a handful of days some two and a half years later, when Neil Armstrong carried them to the moon and back.[76]

Now that the tragedy was behind them, unsurprisingly, there were a number of astronauts who admitted to worrying about what might befall not only Apollo but the overall space program. Walt Cunningham recalled that someone in the astronaut office remarked, "Thank God it happened on the pad." Astronaut Dave Scott also made an analogous statement. "Everyone says how awful. It would have been worse if it had happened in space," he said. Had it been in space, the program likely would have ended.[77]

Similar sentiments were not uncommon. Al Worden worried that Apollo might not see another day. "For all I knew, it might have been the beginning of the end for NASA. Politicians and administrators respond to death in a different way than pilots do. As I went to sleep, I wondered if the loss of Gus and his crew might lead to the cancellation of the entire Apollo program." With the opposition to the program already building up support, a major disaster like this could be all that was needed to end it for good.[78]

V

THE INVESTIGATIONS: "STOP THE WITCH HUNT"

Frank Borman had the admiration of nearly everyone in the astronaut office. Few would say they loved him, for he could be a harsh, no-nonsense kind of guy, but they certainly respected him as both a man and a pilot. He could always be counted on to get the job done. To Deke Slayton he was "tenacious." Anyone who could endure two solid weeks in a sports car–sized Gemini spacecraft circling the earth more than 200 times had to be. John Young called him an "assertive commander" and "one of the toughest of NASA's astronauts." He "was very solemn," wrote Gene Cernan, "and had leadership stamped all over him. Competence was never a question with Frank, because he operated on a higher level than most of us. He wasn't really one of the guys, and somewhat holier-than-thou, but Frank was born to lead." Yet with NASA in a serious situation, a tenacious, tough, holier-than-thou leader was just what the agency needed to help investigate what was then the worst disaster in the nation's space program.[1]

The Apollo 1 tragedy, and the deaths of Gus Grissom, Ed White, and Roger Chaffee, created a watershed moment for NASA. "The history of Apollo is divided into two eras, Before the Fire and After the Fire," wrote Charles Murray and Catherine Cox. And now the space agency had to pick up the pieces and convince the nation that the program was worth saving.[2]

One who did believe it was worth saving was in the man in the Oval Office. "We had come a long way before this catastrophe—not in time, but in accomplishment. I knew every mile of the road we had traveled," wrote the president of the United States. Lyndon Johnson felt a great sense of personal loss with Apollo 1. He had helped create the space program, and it was as near and dear to him as any other. "I have ridden on every flight, and I doubt that any human being could be as concerned or troubled until splashdown as I am or have been," he once said. He was as eager as anyone to get to the bottom of the tragedy and answer every question so that lessons could be learned and the program could move forward with the tragedy behind them. And he had questions of his own. "Demanding questions followed in the wake of the tragedy: What had happened? Could the fire have been prevented? Were we going ahead too fast? Were we pushing too hard? Beyond those questions, I wondered how the nation would regard its space adventure now that disaster had fallen," he said.[3]

President Johnson, though, was wise enough to know the public might conclude that the trip was unnecessary and the cost unrealistic. But LBJ reminded people every chance he got that the whole of the program would cost each citizen about $120 over a nine-year period. Most Americans spend more than that in one year on various vices, like alcohol and tobacco products, he said.[4]

Rocket genius Wernher von Braun did his part to keep the public mind focused on the goal of landing on the moon. "When Charles Lindbergh made his famous first flight to Paris, I do not think that anyone believed that his sole purpose was simply to get to Paris," he said. "His purpose was to demonstrate the feasibility of transoceanic air travel. He had that farsightedness to realize that the best way to demonstrate his point to the world was to select a target familiar to everyone. In the Apollo program, the moon is our Paris."[5]

But, most importantly, despite momentarily slipping public opinion, the astronauts themselves, the ones taking all the risks, wanted the program to continue. The tragedy had hit the astronauts and others in the Apollo program the hardest, especially those who knew Gus, Ed, and Roger the best. Slayton called January 27, 1967, "a bad day. The worst I ever had." Chris Kraft, Director of Flight Operations, admitted that he didn't know "how I survived it." For the three men who would fly on Apollo 11 and be the first to land on the moon in 1969, it was "a bad memory" and "the worst day of all in the Apollo program." And for John Young, who flew with Gus on Gemini 3, it was "NASA's Black Friday," he wrote in 2017. "I swallow hard when I think back to the Apollo 1 fire and the deaths of our buddies.... What those good men experienced was horribly unnecessary."[6]

In the end, the program would continue, helped along by the astronauts who most Americans saw as heroes, von Braun, and "thanks mainly to the strong backing and clout of Lyndon Johnson."[7]

The disaster, though, was certainly preventable. As John Young wrote years later, a "couple of stupid mistakes had cost our three friends their lives." And NASA wasted no time in trying to figure out those mistakes, for the program would never move forward unless a full, honest examination revealed the missteps and offered serious, yet practical solutions. If the space agency failed in this crucial step, everything would be in jeopardy. "I was aware that if we couldn't come up with some right answers, the program might not survive," wrote Walt Cunningham. "The climate was ripe for an emotional backlash against the space effort. The accident would have been even more tragic if the work the crew started had not been allowed to continue and they died for nothing."[8]

Jim Webb understood that too. Within hours of the fire, he initiated the investigation phase of the catastrophe. He delegated authority to Deputy Administrator Bob Seamans to set up an official Apollo 204 Review Board. Understanding the political implications of what he was proposing, Webb went the day after the fire, Saturday morning, to see President Johnson at the White House. When he walked into the residence,

LBJ was still in his pajamas. The issue was whether NASA should investigate the disaster, which was Webb's idea with the review roard, or if an independent commission was needed. Webb obviously favored his own approach and was meeting with Johnson to push his viewpoint, even as the president's science advisor sought an independent committee. Webb laid out his case to LBJ. An independent board might be open-ended, giving the opponents of the program ample time and ammunition to threaten the agency with budget cuts and delays in landing on the moon. But Webb also brought up something that was sure to get Johnson's attention. With an uncontrollable independent investigation, seedy details of political corruption by one of LBJ's former Senate aides over Apollo contracts, which Webb knew all about, would surely come out in the press, giving the president headaches as 1968 approached. If he wanted another term in the White House, a political mess of that magnitude would be the last thing he needed. Johnson understood perfectly. With a handshake, he told Webb, "I want you to handle the investigation."[9]

The board, set up by Seamans with the help of George Mueller, would be chaired by Floyd "Tommy" Thompson, the director of the Langley Research Center. They also brought in experts from various fields to help with the investigation—chemists, explosive experts, safety specialists, and technicians from North American Aviation and other aerospace firms. The board itself consisted of nine members, but overall about 1,500 people were directly involved with the investigation—600 from various federal agencies and 900 from private companies and universities.[10]

Seamans and Mueller also decided that an astronaut needed to be one of the nine members, because the other astronauts "would want assurances from one of their own before proceeding with space missions." They agreed that it should be either Jim McDivitt or Borman. But because McDivitt was deep into work on the lunar lander, and Slayton didn't want him to lose any time on such an important assignment, Borman was the man. And a good one, who had ample experience in spaceflight on Gemini 7 and had been with the space agency for five years. He proved to be "an excellent choice," Seamans said. To Tom Stafford, Borman was known for his "independence and honesty. He wasn't going to let anybody snow him." And that was why he was chosen.[11]

The day of the fire, Borman had not been in Houston or down at the Cape. He had taken his family for a rare weekend trip to a lake-side cabin near Huntsville, Texas, which was owned by some friends, Jim and Margaret Elkins. It was an unusual chance at a few days of rest and relaxation. The telephone number at the cabin was unlisted, so Borman felt at ease knowing no one could find him, at least not very easily. On that Friday evening, as the friends were sitting down to enjoy a nice dinner, there was a knock at the front door. Elkins answered it and there stood a Texas Ranger asking to see Colonel Borman. Frank got up from the table to see what the law enforcement officer could possibly want with him.

"Sorry to disturb you, sir, but the Manned Space Center in Houston has been trying to reach you. There's been some kind of emergency and you're to phone them immediately," the Ranger said. Without haste, Borman called Slayton and received the tragic news of the fire.

"We're organizing an investigation committee, and you're on it," Slayton told him. "Get to the Cape as soon as possible."[12]

Cutting his weekend very short, and after stopping in Houston to see Pat White and straighten out the funeral arrangements for Ed, Borman headed to the Kennedy Space Center at the Cape the next morning to begin what would be an arduous investigation. Key personnel met that night to begin preliminary discussions, then went to a bar to drink away their sorrows. "I was no stranger to violent death and its aftermath, having occasionally been emotionally and professionally close to a victim or victims. But nothing I had witnessed in the past came close to what I saw at Cape Kennedy. The giant complex was in a mass state of shock. Three superbly trained pilots had died, trapped during a supposedly routine ground test," Borman wrote. "I don't think the grief would have been any less if they had perished in space, but at least it would have been more logical and half-expected. The fact that three had died in a motionless spacecraft parked safely on a launch pad was totally illogical, and this was what spawned guilt as well as grief."[13]

The next morning, the real work began when Borman went to inspect the charred spacecraft. The spacecraft was brought to a hangar at Kennedy Space Center where it could be examined from every angle. A make-shift floor was added so that investigators could get into the spacecraft, take photographs, and sift through the remains without compromising the evidence. "As the first one who entered it, I can't even begin to describe that chamber of horror. To me, the interior of a spacecraft had always provided a reassuring sight of gleaming instruments and spotless equipment, creating the illusion of total indestructability. AS-204 was a fire-blackened charnel house, a charred shell that wasn't even a recognizable facsimile of a spacecraft," Borman said. "Hour after hour, I'd sit in the charred cabin—for a long time, I was the only one allowed to enter. I went in first to catalogue and inspect the switches, trying to unearth some unknown flaw in the electrical system. Over and over again I examined the hatch, and finally concluded that if Ed White couldn't open it, nobody could have; its design had frustrated even the strongest of the astronauts."[14]

The Apollo 1 backup crew—Wally Schirra, Donn Eisele, and Walt Cunningham—would be instrumental in assisting the review board, since they had worked closely with Gus, Ed, and Roger and had experience with Spacecraft 012. At the moment, though, Schirra was in Houston "smoldering over the deaths and railing against the inadequacies of a dumb machine built by contractors who should have known better," said Cunningham, who, along with Eisele, was given a role with the review board, although they were not official members of it. They, like other astronauts, were assigned to specific investigative panels. Eisele and Cunningham were tasked with listening to the audio tapes from inside the spacecraft. It was a tough job. "Did you ever listen to your friends scream in panic, then agony, as they fry to death?" Eisele asked rhetorically in his memoir. For Cunningham, hearing the audio "left a sick feeling at the pit of my stomach...because of the horrifying way they had to die."[15]

Eisele also got a chance later on to crawl inside the burned-out spacecraft to help investigators determine the cause of the fire. Like Borman, he was horrified by what he saw. There were "bits and pieces of charred and melted material. The acrid smell of burned plastic, paint, and nylon was overpowering. There were little piles of debris all over the floor and the crew couches. The side walls and the instrument panels were charred, discolored, and warped from the heat. The mess in the cabin and the screaming on the voice tapes gave me nightmares at first. But after a while the dreams went away, along with the knot in my stomach."[16]

Eventually the capsule was disassembled piece by piece. Spacecraft 014, an identical Block I that was designated for the original Apollo 2 repeat mission before it was scrubbed, was shipped from Downey to Kennedy so that it could be taken apart side-by-side with 012. Borman was the board member who oversaw this enormously critical task. The answers to the fire were in the remains of the capsule. According to Slayton, Spacecraft 017, which was already at the Cape and intended for an unmanned test with a Saturn V later that year, was also meticulously inspected, "which turned up nothing but problems. This was bad for all the parties involved: NASA had accepted 017 as it was. It was already in the process of being mated to the Saturn V." Upon inspection of 017, they found "sloppy wiring," many of which were actually skinned, and a total of "fourteen hundred discrepancies." Slayton later wrote that he was "surprised at some of the sloppy workmanship." It seemed like every Block I capsule coming out of the North American Downey plant was in poor shape.[17]

■   ■   ■

Unsurprisingly, NASA was heavily criticized for launching its own investigation, essentially a case of a government agency investigating itself, since all nine members of the review board came from NASA, even though it did have the approval of President Johnson, which was hardly satisfying to critics of the process. One writer for the *Washington Star* called it a "shabby farce." But such an arrangement did have a positive benefit. By keeping the investigation "entirely in-house," wrote John Young, NASA kept it from turning into a "media circus," which is what happened with the Challenger investigation in 1986. Thompson, Borman, and the rest of the board could look into the causes of the fire without facing constant media attention and, hopefully at least, little to no political pressure. But Congress just couldn't help itself. Politicians, especially those who held a critical view of NASA and the space program, were itching to conduct hearings on the fire and showcase their concern before the nation. Senate hearings were scheduled for February. The House hearings, led by Texas Congressman Olin Teague, a great friend of NASA, would be delayed to give the board time to come up with some answers. Before it was over, Borman would be called to testify on several occasions in front of the space committees of both chambers, along with several other astronauts, and the upper echelon of NASA, as well as North American Aviation.[18]

The review board looked at everything from every angle, and identified six conditions that led directly to the tragedy: "A sealed cabin, pressurized with an oxygen atmosphere. An extensive distribution of combustible materials in the cabin. Vulnerable wiring carrying spacecraft power. Vulnerable plumbing carrying a combustible and corrosive coolant. Inadequate provisions for the crew to escape. Inadequate provisions for rescue and medical assistance."[19]

But one issue that the board tried to determine but couldn't was a definite cause of the fire, if that was even possible, given the incredible destruction of the inside of the spacecraft. The official report stated that the fire "was most probably brought about by some minor malfunction or failure of equipment or wire insulation. This failure, which most likely will never be positively identified, initiated a sequence of events that culminated in the conflagration." But the board was able to determine what did not cause the fire. "Electrostatic discharge, spontaneous combustion of flammable material, mechanically produced heat by machinery and heat from the impact of a struck object have been eliminated as reasonable possibilities of ignition of the fire."[20]

For Borman and the rest of the team, "the area of suspicion from the very beginning was the environmental control system," the same unit that had been replaced a number of times. The unit was used to control the temperature inside the spacecraft. With so much instrumentation, it could get quite warm so the ECS pumped ethylene glycol, similar to antifreeze in a car, along with water, through plumbing to "extract heat from the cabin atmosphere" to cool the cabin for the astronauts. But the ECS had been nothing but trouble with constant leaks of glycol. In fact, there were "a number of leaks in solder joints" in every Block I spacecraft. These leaks could cause corrosion of various electrical connections. If glycol leaked on wiring and dried, it left a residue that was both electrically conductive and combustible. During a test in Houston, several ounces of glycol caused a bad fire that burned up a mannequin in a spacesuit. In fact, on the day of the fire, when the crew of Stafford, Young, and Cernan were at Downey doing tests in another Block I spacecraft, they noticed glycol dripping onto the floor from the ECS.[21]

During the investigation, small amounts of glycol were found in the capsule, which could have been caused by the fire burning through the lines. There was no evidence that a glycol leak caused the fire but it certainly could have contributed to its rapid spread. The same company—AiResearch of the Garrett Corporation—had built the ECS for Gemini, and it had worked very well. But now the Apollo system was rife with problems and the company was later "singled out for an especially poor performance." In fact, it had already logged 200 failures. Another problematic area was communication between the two companies. Garrett had offices in a Los Angeles building in close proximity to North American's offices, but they weren't speaking to each other. Soon after the fire, John Young asked that they start talking in order to get the system fixed. NASA determined that there was no other substitute system that could be used to cool the spacecraft, so it directed North American to make sure the leaks were stopped in all future spacecraft.[22]

Since the suspicions of the ECS were answered, the cause had to be electrical. One contributing problem was the amount of highly

flammable material that was inside the spacecraft, such as nylon net-
ting and Velcro, of which there were 5,000 square inches. There should
have been about one-tenth that amount. One North American engineer
said the inside of the spacecraft was "wall-to-wall Velcro." One of
NASA's own, John Dietz, who worked on the Apollo project at the
Manned Spacecraft Center in Houston, had warned for months before
the fire that there was too much flammable material in the cabin,
including spongy foam material laden with urethane. The system, said
Dietz, was one "of carelessness and bureaucratic indifference that
invited disaster." But this had never been looked at as much of a hazard, even in a 100 percent oxygen environment, because there wouldn't
be an ignition source within the capsule, or so it was thought. "This
was wrong assumption number one," Borman wrote, that "the mur-
derous fire couldn't have started by itself, so there had been an ignition
source." And that source was probably the wiring. The board noted
in the official report that during tests "in a 16.5 psi oxygen atmosphere," generated "sparks blown from [electrical] arcs can ignite
combustible material several inches from the arc."[23]

The board "never did determine definitely what had caused the
spark, although we would have bet our combined bank accounts that it
originated in a bundle of electrical wiring in front of Grissom's feet. It
had to have been some kind of short circuit, because we did have evidence
of a bright arc that appeared in this one area just before fire swept
through the cabin," Borman later wrote.[24]

Others were more sure of the cause. Writing years later, John Young
believed that the "immediate cause of the launch pad fire was damn
mundane, even trivial. An electrical wire on the floor of the spacecraft's
lower equipment bay had become frayed, probably due to the procession
of technicians in and out of the spacecraft in the days before the test."
And many of the technicians were repairing and replacing the ECS. "The
wiring in both the capsules that North American had manufactured so
far was very bad. Big wire bundles lay against the aluminum stringers
with no support. With the g-loading we'd be getting, such unsupported

bundles could easily fail. The bundles were far larger than they should have been. We saw many instances of wiring where insulation was already frayed. In pure oxygen this was not good."[25]

In the official report, the board noted that there were "numerous examples in the wiring of poor installation, design, and workmanship." Wiring could likely have been frayed "underneath the lithium hydroxide access door" because insulation covering power cables "passing under this door was potentially vulnerable to abrasion from the corner along the lower edge of the door," but the fire damage in that area was "so extensive" that it could not be conclusively proven. "It is most probable," the official report noted, "that the fire was initiated by an electric arc either in this location or in some other region near the Environmental Control Unit."[26]

In his congressional testimony, Borman disclosed to Senator Carl Curtis of Nebraska that there were "defects of workmanship" throughout the spacecraft. The official report even noted that a socket wrench had been left in the capsule, which was discovered during the disassembly process. The spacecraft systems might have been "functioning normally" on the most recent test but that doesn't mean the entire craft was free of all possible ignition sources. As Borman stated, there were "defects in the design of the wire bundles, their routing, their construction, and in my opinion, a basic deficiency in the wiring, in the harnesses, that distribute electrical energy." Even though the review board did not locate a single ignition source, it is quite obvious that there were a number of wires that remained defective. There's simply no way to be certain that in thirty miles of wire one or two defective sections could have been overlooked.[27]

Worse still, many of the astronauts had known there were problems with the wiring and sloppy workmanship on the part of North American. Just a few weeks before the fire, John Young asked Gus why he had not yet said anything about the bad wiring, because it "was nothing like the wiring we had in our Gemini spacecraft with respect to support, bundle amounts, and overall quality of every bundle." But Gus, who had complained about so much with the spacecraft, felt that pushing too hard on the issue might be a bad idea. "If I say anything about it, they'll fire me," he said.[28]

Young felt a sense of guilt about not saying something himself, for he "knew a lot about proper electrical wiring in a flying machine." With a 100 percent oxygen atmosphere, Young understood, "If anything set off even a tiny spark, the results would be fatal." There were close calls. The day before the fire, Dave Scott was training in a Block I spacecraft in Downey and, while in a pressure suit filled with 100 percent oxygen, got a "pretty strong electrical shock." He was "very luck he didn't get electrocuted, burnt to death." While Young was working on Gemini, a similar ground test in 100 percent oxygen "had caused my knees to shake. Maybe if I had asked the right engineers at JSC [Johnson Space Center in Houston] to look into it, Gus, Ed, and Roger would still be here today." But perhaps not, he ultimately concluded. "Today it's hard to comprehend just how ignorant some of our early spacecraft management teams actually were. Even if the bad wiring and the danger of testing in 100 percent oxygen had been laid out in front of them plain as day, NASA might not have changed course, short of a catastrophe."[29]

Another issue confronted by the review board, a second of Borman's assumptions, was the pressurization of the spacecraft. In space, the cabin would be pressurized at 5.5 psi. But during the test, the spacecraft was basically on the ground, sitting atop the Saturn IB, which was over 200 feet above the pad, where the outside atmospheric pressure was 14.7 psi, so there was far greater pressure inside the spacecraft, at 16.7, for the test. And that, coupled with the 100 percent oxygen environment, made for a potentially deadly combination. "That is an extremely dangerous environment," wrote Borman, "the equivalent of sitting on a live bomb, waiting for someone to light the fuse."[30]

But no one had much thought that a fire on the ground during a routine test was remotely possible, not in Mercury or Gemini, when 100 percent oxygen under pressure was used in test after test, the very kind that had given Young the shakes. Walt Cunningham admitted that the "possibilities of a fire on the ground had been given precious little thought. The cabin didn't even have a fire extinguisher." Yet NASA had known of the risks, for a number of scientists had studied the issue in the years before the fire and repeatedly warned that such a spacecraft environment was dangerous.[31]

In 1964, Dr. Emmanuel Roth compiled a four-part report for the Lovelace Foundation for Medical Education and Research entitled "The Selection of Space-Cabin Atmosphere." His conclusion was that flammable, combustible materials in 100 percent oxygen would burn and burn violently. He also concluded that non-flammable items would also ignite and burn in a similar manner.[32]

In the same year, Dr. Frank J. Hendel, who worked at North American Aviation as a scientist with the Apollo Space Sciences and Systems division, penned an article for the *Journal of Spacecraft and Rockets*. In his article, "Gaseous Environment during Space Missions," he wrote, "Pure oxygen at five pounds per square inch of pressure presents a fire hazard which is especially great on the launching pad.... Even a small fire creates toxic products of combustion," while, at the same time, "no fire-fighting methods have yet been developed that can cope with a fire in pure oxygen."[33]

Ronald G. Newman, a contributing editor of *Space/Aeronautics*, wrote a review of *Factors in the Operation of Manned Space Chambers*, a conference of the American Society for Testing and Materials held in Seattle in November 1965. In his review, Newman wrote, "With reliability figures and flight schedules as they are, the odds are that the first casualty in space will occur on the ground." At the same time, in an editorial for the magazine, in which Newman was likely a contributor, the editors wrote, "By now, NASA and other involved agencies are well aware that a regularly updated, progressive set of recommended practices—engineering, medical and procedural—for repressurization schedules and atmospheres, medical monitoring, safety rescue and so on, would be welcome in the community."[34]

Aside from scientific opinion, NASA also had a number of examples of actual incidents that had occurred in the U.S. military around the same time. There were oxygen fires at Brooks Air Force Base and the Navy Equipment Diving Unit in Washington, D.C., in 1965, and the Airesearch Facility in Torrance, California in 1964, 1965, and 1966. The latter incident occurred on April 28 in an altitude chamber used to simulate the interior environment of the spacecraft operating in space, with 100 percent oxygen at 5 psi. "This should have been a wake-up call," said NASA engineer Gary W. Johnson, who was on the panel that investigated that particular fire. NASA should have known that a pure 100 percent oxygen environment, coupled with highly flammable material, was a dangerous atmosphere to have in the spacecraft, especially one sitting on the ground during a test. The Soviets used a two-gas system for those very reasons, but NASA had chosen not to.[35]

Joe Shea also knew of these fire risks with 100 percent oxygen. As early as March 1965, Shea and his team had wrestled over the issue of a two-gas system—a mixture of oxygen and nitrogen—which would be far less flammable, but concluded that a pure oxygen environment was preferable because it was safer, less complex, and lighter than a two-gas system. The idea seemed to be that if there were no flammable materials in the spacecraft, and no obvious ignition source, then the likelihood of fire was remote. The previous reports aside, Shea had been personally warned at least twice about the possibility of fire with pure oxygen. Hilliard W. Paige of General Electric and Dr. Charles Berry, NASA's medical director, had made their opinions clear to him. Paige's warning came in September 1966, just four months before the plugs-out test. Because of the use of 100 percent oxygen, the "first fire in a spacecraft may well be fatal," Paige wrote to Shea. After looking into it, Shea wrote Paige seven weeks later. "We think we have enough margin to keep the fire from starting—if one ever does, we do have problems."[36]

According to Lawrence Korb, an engineer for North American, the company proposed a two-gas system in its initial bid for the Apollo contract but "NASA had strong objections because it required two tanks, separate regulators and some sort of sensing device to maintain the right mixture so astronauts would not pass out. We, led by Dr. Toby Freeman, argued vehemently against NASA's Max Faget. NASA told us the decision was final: pure oxygen."[37]

To be fair, a dual gas system wasn't as simple as pumping oxygen and nitrogen into the spacecraft, writes space historian Amy Shira Teitel. "Balancing the gases demanded North American invent some way of measuring the mixture continually, adjusting the mix of gases constantly with every change. If that system failed, the crew might lose consciousness before realizing there was a problem. A pure oxygen system wouldn't just be lighter, it would be far simpler; all the crew would need was a simple pressure sensor to ensure the cabin was adequately pressurized. This rationale was enough to have NASA change the Apollo crew cabin from a mixed gas to a pure oxygen environment."[38]

After the fire, Webb said that the decision was made to use pure oxygen after "the utmost responsible examination of all alternatives." A NASA spokesman clarified his remarks by adding, "At the time this decision was made, the available evidence indicated that the fire hazard of pure oxygen was no greater than that presented by a two-gas atmosphere." But the warnings NASA had been given contradict both of those statements.[39]

These potential problems did not seem apparent at the time, though. George Mueller testified before the Senate that previous experience had taught NASA "that the possibility of a fire in the spacecraft cabin was remote." Yet they were blatantly obvious years later. "No one had taken the time to consider that an electrical spark in the spacecraft, while unlikely, would be disastrous," recalled Al Worden, who would fly to the moon as command module pilot on Apollo 15. "For three dangerous ingredients to come together like this—flammable material, sparking wires, and pressurized pure oxygen—a lot of details must have been overlooked by a lot of people, and not just those at North American. These details were probably missed because, until January of 1967, everything had worked fine, so we were overconfident. No one wanted to stop and think about potential problems. In this instance, it killed three men." Astronaut candidate Dr. Fred Kelly, who was a Navy pilot and flight surgeon and led the medical team during the investigation, concurred. "With 20/20 hindsight anyone off the street should see that a spacecraft with 72.5 pounds of flammable material, miles of unprotected wiring and 47 possible ignition points in an atmosphere of 100 percent oxygen at 16.7 psi was an accident waiting to happen."[40]

"The fire was absolutely unexpected to us," said Glynn Lunney, a NASA flight director. "In retrospect, it was a risk we should have defensed in a design sense. We should have defensed it, but we didn't." For Jim Lovell, NASA "should've had plenty of warning ahead of time. We flew Mercury. We flew Gemini. All those spacecraft were tested on the ground by pumping in pure oxygen," he said. "Now for all of the Gemini flights, the Mercury flights, no problems. But Apollo, probably maybe due to the rush they were trying to do…they pumped it up—we should've realized that you didn't have to put pure oxygen into that spacecraft. You should've put a mixture in it anyway of oxygen and nitrogen, because anything will burn in pure oxygen, as found out in the Apollo 1 fire." John Young felt the same way, writing that the "use of 100 percent oxygen inside that spacecraft during a ground test was a fatal mistake, one that we all should've caught months, if not years, earlier."[41]

In 1987, twenty years after the fire, North American Aviation's former president, Lee Atwood, wrote a letter to astronaut Wally Schirra about the tragedy. "If the question had been properly put," he said, "that is, did you know that the astronauts are being locked in with all that electrical machinery, and the spacecraft is being inflated to 16.7 pounds per square inch with pure oxygen? I believe a whistle would have been blown."[42]

The other major problem was the capsule's hatch, which Ed could not get open so the crew could egress the spacecraft. Several astronauts and engineers had petitioned Joe Shea two years before the fire to install a simple, one-piece hatch that opened outward rather than the two-piece hatch designed by North American Aviation. "The hatch design was a disaster," wrote Al Worden, because "it sealed from the inside, so the greater the interior pressure, the tighter the seal," which explains why Ed probably would have had a hard time wrenching it open even if he had had the time. It was cumbersome to handle, noted Donn Eisele, because it sealed from the inside and had greater pressure pushing against the seals "so hard you couldn't pull it loose with elephants. I remember looking at Ed White's finger scrapings in the melted goo on the inner hatch surface. He must have pawed frantically, trying to get the hatch open, just seconds before he expired from searing hot gases that suffocated him and burned out his lungs." The design of the hatch, said Worden, was "an example of a system designed by nonpilots: safety was not the most important factor." It was described by several astronauts as "a monster," "pretty damned heavy," and "a brute of a thing."[43]

According to Eisele, Shea turned down the proposal for a new re-designed hatch. "Too much money and not enough time," he told the astronauts. "It's a 'crew comfort' item. You guys are just too lazy to wrestle with the hatch we've got. What the hell, it works, doesn't it?" But it didn't work and the crew perished trying to get it open. As the official report noted, "The crew was never capable of effecting emergency egress because of the pressurization before rupture and their loss of consciousness soon after rupture." In an exchange with Senator Walter Mondale of Minnesota, Mueller discussed the hatch issue, noting that the fire increased the pressure to 36 psi, making it impossible for the crew to open the inner hatch to the inside. But Mueller also noted, once the rupture occurred, "the pressure dropped immediately down again." The problem, though, was time. The crew was likely dead, or at the very least nearly unconscious, by the time the rupture occurred. There was a way for the astronauts to relieve the pressure from the inside but there most likely wasn't enough time. The fire quickly became an inferno and killed the crew within seconds. In reply to another question from Mondale, Mueller also noted that there was no way for the technicians on the outside to relieve the pressure either. The astronauts were effectively in a deathtrap.[44]

The investigations produced a lot of information from the review board and that which came out of the congressional hearings. The ultimate cause would be debated for years. But one man who worked on the investigation believes he knows how it happened. Scott Simpkinson, the engineer who oversaw the actual disassembling of the burned-out spacecraft and edited the official report when it was released, believed that the cause of the fire was the metal door to the compartment for the lithium hydroxide canisters, which contained a sharp edge. The official report listed this particular cause as a possibility but Simpkinson believed it to be fact. As he told it, the opening and shutting of the door, which was positioned so that it opened into the environmental control unit, caused wiring to be exposed, which likely shorted and sparked during the test. Leaks from the ECU, causing the presence of flammable glycol fumes, created the fire.[45]

Simpkinson also suggested that perhaps Gus was responsible for the scuffed wire by opening that door during the test, which caused the electrical arc that started the fire. This particular possibility was raised during the congressional hearings in the House by one of North American Aviation's executives, John McCarthy, a charge that Slayton called "pathetic." As Erik Bergaust wrote in *Murder on Pad 34*:

> It took a moment for the full implication to sink in—that Grissom's own carelessness caused his and his colleagues' deaths. The suggestion immediately recalled another incident during Grissom's first space flight in July, 1961, when the hatch of his *Liberty Bell 7* Mercury spacecraft blew off in the water after landing. The spacecraft sank in the ocean, never to be retrieved, and Grissom nearly drowned. McDonnell Aircraft Co., the spacecraft manufacturer, faced with a charge that something had gone wrong with explosive charges on the hatch, tried to show that Grissom, supposedly in panic, had blown the hatch himself.[46]

But *Liberty Bell 7* was not on the minds of the assembled members of Congress that day; the tarnishing of a dead national hero was.

And that insinuation did not go well. It greatly angered Slayton. "Given Gus's history with the Mercury hatch, of course, this just played right into the image of Gus as some sort of screwup. It really pissed me off, and I wasn't the only one, especially because there were no grounds for the story—it was pure speculation, not to mention physically impossible—and because Gus wasn't around to defend himself the way he was on the *Liberty Bell* incident."[47]

Congressman William F. Ryan of New York also took great offense at McCarthy's suggestion. "I take exception to your trying tonight to place the blame on someone who is not here to speak for himself." Though McCarthy tried to excuse his statement as a simple "hypothesis," most of the representatives were not having it. It was as if North American was subtly trying to shift the blame for their sloppy work onto a possible mistake by Gus, so as to somehow absolve the company of any responsibility. Congressman James Fulton of Pennsylvania was equally offended. "As a friend of Gus Grissom, I feel I must clean the record. In my opinion as an attorney of some years' experience there is no scintilla of evidence based upon any physical condition noted after the fire upon which to base a prima facie presumption of probable cause that astronaut Gus Grissom in any moment kicked or disturbed with his foot any equipment that had any bearing upon, or contributed to, the cause of the accident."[48]

To make his case more fully, Fulton went right to those whose opinions garnered the most respect, the astronauts at the hearing. He asked Borman for a professional opinion as to the possibility of such a "hypothesis." Borman answered him succinctly. "We found no evidence to support the thesis that Gus, or any of the crew members kicked the wire that ignited the flammables." A definitive answer by an astronaut of Borman's stature was usually enough to slam the door shut on any argument during the hearings. It wouldn't be the last time Frank Borman put Congress in their place.[49]

■   ■   ■

The exhausting, and oftentimes emotional, work of the review board took its toll on those involved in the investigation. "Sometimes it was hard to stay emotionally detached from our task. It got to us in different ways. I began to get progressively angrier at what we gradually unearthed—sloppy planning and supervision on NASA's part and some shamefully inadequate design and test work by North American," wrote Frank Borman. It didn't help matters that the investigation turned into "a feeding frenzy," said Deke Slayton. "Everybody was going around pointing fingers at everybody else. The strain was incredible." That was true even at Mission Control. "We didn't get over it for well over a year," said John Boynton. "Those guys [Gus, Ed, Roger] were so helpless. It was the worst tragedy in NASA's history, including *Challenger* and *Columbia*, because it was on the ground. It should have been prevented."[50]

For Bob Seamans, NASA's deputy administrator, the "trouble was not that our people didn't care enough about the fire; they cared too much. Key people from Houston would fly up to Washington to testify and literally sob all the way on the plane." Jim Webb began suffering migraine headaches. "He took all of this personally. He became terribly tense," noted Seamans. "I had the feeling I was dealing with somebody who could explode at any moment."[51]

Borman and Seamans both knew that there were a few involved with the investigation who became more and more depressed about the investigation's finding and began to take prescription drugs to cope— downers to deal with the pain and anguish, then uppers to be able to function the next day at work, which caused some eventual breakdowns, including of John Yardley, who worked as a consultant to the board from McDonnell and had been working with NASA since Mercury. A few even had to be committed and got psychiatric treatment, one of whom was carried to a facility in a straitjacket, while others turned to the bottle.[52]

But perhaps the hardest hit was Joe Shea. "North American's skirts were far from clean, and neither were NASA's. We became painfully aware that part of the trouble lay in the management of Project Apollo itself, and that put the onus on poor Joe Shea's doorstep," Borman wrote. "He was unquestionably an excellent scientist, but also a poor administrator who had simply let North American's design mistakes pile up like unnoticed garbage."[53]

Most agreed with Deke Slayton that Shea was "a brilliant guy" who was "on his way to being famous." *Time* magazine was set to run a cover story on Shea, due out the week of Apollo 1's flight in late February. But the fire ended that story. "Joe took the fire personally," Slayton wrote. "It wasn't that he collapsed; just the opposite. He went on overdrive, as if he could personally redesign and rebuild the spacecraft so it would never fail again." Bob Seamans also knew Shea well and worked closely with him. He "was the most affected," Seamans later wrote. "I could see he was extremely upset." Shea became an obsessed workaholic, sleeping just four to five hours a night, determined to get to the bottom of the tragedy.[54]

But the exhaustive work had other effects as well. Chris Kraft was told that "both Shea and Yardley were getting stupefied drunk every night." Then, within weeks, "it came to a head" during a meeting in Houston. "Joe Shea got up and started calmly with a report on the state of the investigation. But within a minute, he was rambling, and in another thirty seconds he was incoherent. I looked at him and saw my father in the grip of dementia praecox. It was horrifying," Kraft thought. "Joe had become a friend. Now he was falling apart in front of me." General Sam Phillips, who was running the meeting, intervened, thanked him for his work, and asked him to sit down. "That was the end of Joe Shea."[55]

Part of what was affecting Shea was a sense of guilt. Gus had wanted him inside the spacecraft with the crew so he could get a first-hand look at the problems. He refused mainly because it would have been almost impossible to get him hooked up with a communication headset so he could listen in on the conversations with test conductors. But had he been in the capsule he might could have acted to stop the disaster from unfolding. "I'm probably 70 percent convinced...that had I been down there I probably would have seen it [the fire start], I think I could have grabbed it; I probably would have burned my hands or something—and I could have smothered it enough, so that it never would have spread." At another time, he told a reporter that he wished he had been inside during the fire and "gone with those guys." Who knows if it were even possible to stop the fire in such an oxygen-rich environment, but the guilt was clearly eating at him.[56]

Despite his hard work and intense determination, and with Jim Webb's support, Shea was moved out of his stressful job in Houston to NASA's headquarters in Washington. Eventually he left the space agency for private enterprise and later a professorship at MIT.[57]

■   ■   ■

Congressional hearings by both houses of Congress lasted for months, as representatives of the people and keepers of the public treasury, of which bounty NASA, and by extension North American Aviation, had received a large share, wanted their say. Webb, Mueller, and other NASA executives, as well as top people at North American all had to face the congressional music and the persistence of the media. For NASA, Bob Seamans was tasked with keeping the press informed about what was happening during the hearings. And he knew how huge of a deal it was the first time he walked into the hearing room, noting that he "had never before seen more coaxial cable strewn around a corridor."[58]

Most of the Apollo 1 hearings, like any congressional activity, especially one encompassing thousands of pages of testimony and exhibits, could bore the paint off the wall. But there were some tense exchanges and moments of rhetorical fireworks. Throughout the hearings, both NASA and North American kept a united front, equally accepting blame for the disaster, at least to start with. But as the heat increased, the loyalty sputtered. North American bosses blamed NASA for a flood of modifications, some in late 1966 when the spacecraft had already been delivered to the Cape, which, they claimed, added 70 percent more wiring to the capsule. Seamans accused North American of not showing "sufficient dedication to the engineering design or workmanship on the job." Watching the drama unfold, one *New York Times* reporter wrote, "Each partner is now reaching into the file cabinets for carefully preserved records that would tend to absolve itself and shift the blame to others." But the biggest bombshell that came out of the hearings was the revelation of what was known as "The Phillips Report."[59]

General Samuel C. Phillips came on board as the new Apollo Program director at NASA headquarters in Washington in 1964. Borman called him "a straight shooter who hated red tape and procrastinating bureaucrats as much as I did." To *New York Times* space reporter John Noble Wilford, Phillips was a "tall, lean air-force general, who was trained as an electrical engineer, had managed the development of the Minuteman ICBM, an effort with many similarities to Apollo in that advanced technologies and rigid scheduling were guiding factors. Phillips was to emerge as the key decision maker, prodding companies to get their work done on time, monitoring their costs and performance, giving the go-ahead signal for flights." George Mueller, NASA's associate administrator for manned space flight, was concerned about North American's work on the Apollo spacecraft and the second stage of the Saturn V and had requested that Phillips put together a "tiger team" and head to Downey to find out what was going on.[60]

After several days of investigation, which lasted from November 22 to December 6, 1965, Phillips compiled copious notes that formed the basis of the Phillips Report. "As the program progressed," Phillips wrote, "NASA has been forced to accept slippages in key milestone accomplishments, degradation in hardware performance, and increasing costs." The word "slippage" appeared numerous times throughout the report— slippages in schedules and production. Components were months behind schedule and the command and service module still had "significant problems." Quality of production "is not up to NASA required standards." Cost overruns were also a major issue, as North American "has not been able to forecast costs with any reasonable degree of accuracy," Phillips stated. In fact, North American had said that the initial phase of the contract would cost $400 million or so, but by 1967, they had spent $2.2 billion; by 1971, it was $4.4 billion. The team made recommendations to North American and gave notice that they would visit again in March 1966 to see if the recommendations had been implemented.[61]

On December 19, 1965, Phillips had sent a private letter and ten copies of his notes to the president of North American Aviation, Lee Atwood. "I am definitely not satisfied with the progress and outlook of either program," he wrote, referring to both the spacecraft and the Saturn V's second stage, "and am convinced that the right actions now can result in substantial improvement of position in both programs in the relatively near future." Time was of the essence, Phillips noted. "The gravity of the situation compels me to ask that you let me know, by the end of January if possible, the actions you propose to take." The thinly-veiled inference was that the contract might be cancelled if North American didn't take appropriate steps to correct these defects, although that was never explicitly stated.[62]

When Webb appeared before the Senate Space Committee, he was blindsided by revelations of the Phillips Report by Senator Mondale, a liberal Democrat who was critical of NASA and the Apollo program. Bob Seamans was there, alongside Webb, Mueller, and George Low. "Mondale was the junior senator, so he came on last," Seamans said. "And it had been a long, long session, and so by the time he came along— I won't say we were waning, but we'd been through quite a bit already." Then Mondale hit them with a question about the report.[63]

"Is there anything to be learned in comparing the successful operation of the Gemini series with what we have already experienced thus far," Mondale began, "and let me preface this by saying that I have been told, and I would like to have this set straight if I am wrong, that there was a report prepared for NASA by General Phillips, completed in mid or late 1965 which very seriously criticized the operation of the Apollo program for multi-million dollar overruns and for what was regarded as very serious inadequacies in terms of quality control."[64]

Mondale continued, "This report, among other things, was so critical that it recommended the possibility of searching for a second source [another company to build Apollo], and as I am told, recommended Douglas Aircraft.

"Would you comment on that? Is there a Phillips Report?"

No one said a word for a few seconds, sitting in stunned silence. "I had a sense that Mondale wasn't just on a fishing trip. He knew something, and the question was—what did he know?" Seamans wondered.[65]

Breaking the silence, George Mueller spoke first with a long-winded, technical answer. General Phillips had certainly examined a number of contractors and written reports based on his findings, and "not all of them have been completely complimentary," Mueller noted. "I think you will recognize that in any research and development program, we have to pay attention to areas of weakness, both within NASA and among the contractors."

Not satisfied, Mondale asked again. "Is it your testimony that there was no such unusual General Phillips Report? Is that rumor unfounded?"

Mueller's reply was less than truthful to say the least. "I know of no unusual General Phillips report."

Mondale had the strong feeling that NASA "really wanted to cover it up. They didn't want to get into it. They didn't want to disclose it. They didn't want a public discussion of the problems that the space program was into. And so I kept pressing them," he said years later.[66]

"Was there a report in which General Phillips recommended looking for a second source?" Mondale asked Mueller in a follow up question.

"I do not recall such a report," Mueller replied. To that Mondale sought to clarify the issue. "So that, basically, what I have incorporated in my question here by way of what has been reported to me, you regard to be inaccurate?" Mueller did not reply before Webb stepped in with a long response that did not really answer the specific question but did not admit to the existence of the report either.

Speaking to the press afterwards, Seamans felt the need to explain Mueller's answers. "I interjected a few thoughts about the kind of review we normally carry out. I added that I wasn't aware of any specific report, which I wasn't. As far as I knew, there had never been a recommendation to go to another contractor for the spacecraft."[67]

That particular issue might have been true but Webb was not happy with Seamans's statements to the media. He pulled him aside after the conclusion of the hearing. "You don't volunteer information!" he said rather forcefully. "You can't look at these as ordinary hearings, like any we've had in the past. You've got to look at these as legal proceedings. Don't volunteer information unless you're sure. Don't volunteer information, period!" he said. "Don't speculate in front of Congress!"[68]

In his memoir, Seamans tried to defend Mueller's answer. "I knew George meant that there hadn't been a formal, bound report." Perhaps that's true, or perhaps Mueller was centering on Mondale's use of the word "unusual," since reports on contractors were far from unusual. But clearly Mueller had misled Congress; he knew all about the report. On the same day General Phillips wrote to Lee Atwood at North American, December 19, 1965, Mueller also sent a letter of his own, pointing out the findings of the Phillips Report and his own complaints. As it stood then, the flight of Apollo 1, scheduled for October 1966, would have to be delayed, he said. "Late last year, when the Block II Program was defined, your people agreed that they could and would do a better job on Block II engineering," he wrote, yet that "has been neglected" and is "months behind schedule." The Saturn V second stage was no better. "It is hard for me to understand how a company with the background and demonstrated competence of [North American Aviation] could have spent four and a half years and more than half a billion dollars on the S-II Project and not yet have fired a stage with flight systems in operation." Mueller further questioned North American's competence and, like Phillips, notified Atwood that the team would be back in March. For his part, Atwood also lied to Congress when he was asked about his knowledge of the report, even though both Phillips and Mueller had written to him about the findings and sent copies of the "notes" to him.[69]

It's clear that Mueller knew of the report and had to know exactly what Mondale was talking about. Webb likely did as well, although there is no document, or any corroborative evidence, that proves it. When Mondale asked him about the report, Webb also denied knowing about it. He even tried to confuse the issue, asking Mondale if he was "thinking of a report circulated by a former employee of the North American Aviation company recently within the last six weeks." Here Webb was referring to the Baron revelations.[70]

Thomas Baron worked for North American Aviation as an inspector and did not like what he was seeing. By the beginning of January 1967, Baron had been with the company for nearly a year and a half, and in all that time he kept detailed notes on everything he witnessed or heard from other employees, much of it apparently rumor, gossip, and hearsay. Yet his findings totaled some 55 pages of notes. Baron's complaints were "poor workmanship, failure to maintain cleanliness, faulty installation of equipment, improper testing, unauthorized deviations from specifications and instructions, disregard for rules and regulations, lack of communication between Quality Control and engineering organizations and personnel, and poor personnel practices." And much of this was discovered later by the review board. Yet as authors Charles D. Benson and William Barnaby Faherty point out, NASA took notice of Baron's issues and looked into them, formally interviewing him and then briefing members of NASA management on what they had found. North American fired Baron on January 5 but the company also looked into his allegations. "Both NASA and North American Aviation, a historian must conclude, gave far more serious consideration to Baron's complaints than a casual perusal of newspapers during the succeeding weeks" after the fire, write Benson and Faherty. There just wasn't enough time to deal with everything before the test on January 27. Baron testified before the

House investigating committee at the Cape in April, but tragically, a week later, the former North American employee, who had health problems and at least one stint in a psychiatric facility, was killed, along with his wife and step-daughter, when their car was struck by a train. The circumstances of the crash, writes Piers Bizony, "were not inconsistent with a suicide."[71]

But Mondale wasn't buying any of that. "No, the so-called Phillips Report," Mondale retorted, "which came out in mid or late 1965, conducted by NASA on the Apollo program."

"Let us look it up," Webb replied.[72]

In a 1999 interview, thirty-two years after the hearings, Mondale recalled Webb's reaction. "He looked dazed and stunned. And I still don't know if he knew about it. I think he did. Remember, I asked him to produce a copy for the committee and he said, as I recall, that he would certainly look at that but we had to be aware that there was a lot of sensitive information that may not be available. And he didn't give me a yes-or-a-no answer. But at least I had raised the issue. The committee was now aware of it. And the national press was alerted to the possibility that this report existed. I think it was inevitable from that point that it would be produced."[73]

Seamans, though, had worries. "I was very concerned about the Phillips study," he said, and "felt it was the kind of thing that could cause very great trouble." He "thought the way to solve it was to get the facts" to Congress, specifically the committee chairman, Senator Clinton Anderson. Back at his office after the day's hearing, Seamans was finally shown the Phillips Report, and realized how bad it was. "Take it in to the boss," he said. This is, perhaps, the strongest evidence that Webb hadn't seen it.[74]

But decades later, Seamans appeared to change his tune, at least a tiny bit. As deputy administrator, Seamans saw Webb every day and briefed him on everything that was on-going with the program, but years later he seemed puzzled as to why he wouldn't have told him about the Phillips Report. "I'm sure we brought it up at the time," he said. "For the life of me, I can't believe that I didn't keep him informed. But he claimed after the fire that no one had ever told him about these problems," he said, perhaps with a slight bit of skepticism in his old boss's account. "That's one of the things I feel badly about, that at least he felt that he'd been kept out of the loop." For his part, Webb admitted that he had spoken with Seamans about the problems with North American, and if Seamans mentioned the report, "it was casual and not flagged as something important."[75]

One historian, John Logsdon, who worked for NASA and served on the review board for the *Columbia* disaster, believes that Jim Webb did not know about the Phillips Report simply because of George Mueller and how he ran his part of the NASA hierarchy, the Office of Manned Space Flight in Washington. Mueller was a strong character and hard for Webb to control, running his office "as essentially a separate organization, to the point that Webb didn't always know what he was up to. You have to keep in mind that the Phillips Report was originally made for Mueller to see, not Webb."[76]

But it is hard to believe that Webb did not know about something as serious as the Phillips Report. If he truly didn't, it could have been a classic case of plausible deniability. Either that or he was asleep at the wheel, and Jim Webb was never asleep at the wheel, so it was likely the former rather than the latter. No matter the excuse, though, as NASA's administrator, he should have known about it. That was the opinion of *Washington Star* reporter Bill Hines, the same journalist who expressed skepticism at NASA's investigating itself. "To persons familiar with NASA's organization, it is utterly inconceivable that Webb would not have known about such a far-reaching document as the Phillips Report," he wrote. "It can be stated quite without question that this sort of letter simply would not be written, or such a report rendered to the agency's largest contractor, without the knowledge of the agency's administrator."[77]

Whether Webb knew of the report or not was now secondary to the revelation of the report by Mondale at the congressional hearing. Making matters worse, it had to be crystal clear to NASA executives that someone inside the upper management of the space agency had obviously leaked the memo or at least a summary of it to Mondale, or at least told him about it. Mondale admitted in 1999 that he did not have a copy but heard about it "from a source that was very credible." Who could it have been? Such a thought could only have sent shivers down the spines of NASA officials assembled at the hearing.[78]

With the cat out of the bag, Webb and top NASA officials still refused to turn over a copy of the report to Congress, which only led to charges of a cover-up. A copy of the report found its way into the hands of Congressman Ryan, who threatened to release it to the press if NASA did not turn over a full, unredacted copy to Congress. Within two days of Ryan's threat, NASA turned over a copy to Congress. It had taken a month and a half to do so, and during that period of stalling Mondale noted that NASA witnesses were dancing a "semantic waltz" to avoid acknowledging the report.[79]

Though it may sound like a typical bureaucratic cover-up, as some have alleged, the facts suggest something far less dramatic. The Phillips Report was simply an internal report revealing issues that NASA needed to address with North American Aviation. It was never meant for public consumption. In fact, Phillips himself referred to his findings simply as a "set of notes," while Slayton called it "troubleshooting" that was brought about because of the trouble brewing between NASA and North American. To Slayton, North American was "basically acting as though they were the experts on spaceflight while NASA was just a bunch of government people who really didn't know what they were doing." But, as for NASA, Slayton wasn't shy about casting some blame, admitting that the agency "drove the contractors crazy with unrelenting schedule and money pressure, and thousands of last-minute changes in design." Phillips, at Mueller's request, simply wanted to inspect the program and see where the problems were to get them addressed. His "tiger team" went to Downey to see for themselves.[80]

Revealing it to the public at large, or worse to Congress, would only have made Webb's job that much harder, not easier. Mondale had asked

Seamans during the Senate hearing, "Well, shouldn't all of this have been released?" Seamans replied that if "every time we had an internal review, everything was released, good internal reviews would become unachievable. If they are made with the idea that they will be discussed on national TV, they will never contain controversial information." In other words, if everything were open and available, NASA might never get the truth from contractors, which was imperative in an occupation as complex, dangerous, and hazardous as manned spaceflight.[81]

Senator Clinton Anderson, chairing the Senate hearings, made note of the need to maintain at least some modicum of confidentiality. "Notwithstanding that in NASA's judgment the contractor later made significant progress in overcoming the problems, the committee believes it should have been informed of the situation. The committee does not object to the position of the Administrator of NASA, that all details of Government/contractor relationships should not be put in the public domain. However, that position in no way can be used as an argument for not bringing this or other serious situations to the attention of the committee."[82]

At the time, though, there was already growing opposition in Congress and in the general public to the moon program, and a report of that magnitude would have hit like a bomb. Many in Congress would have immediately politicized it, thereby making the situation much worse. It didn't need to be secretly leaked or openly handed over to Congress because NASA was hard at work addressing each issue in the report and by the time of the fire had managed to fix a number of them. By launch day, it is possible that all of the specific issues would have been addressed given the additional month remaining in the schedule. But that may not have prevented the tragedy, because what the fire revealed was that even if everything in the Phillips Report had been taken care of, the Apollo spacecraft still would not have been flightworthy enough to carry out lunar missions, and probably not good enough to carry out much of an earth-orbit flight either. That much was made obvious when the smoke cleared. The fire tore open the secrets locked deep inside the capsule, secrets that no one could have known, and showed just how bad the spacecraft actually was. And that was the "hidden blessing in this disaster," wrote Andrew Chaikin, in that "the wreckage of Apollo 1 was there for the accident board to examine, not a silent tomb circling the earth or drifting in the translunar void." By revealing the problems, the fire made it possible for a redesigned spacecraft that was infinitely safer and capable of flying Americans to the moon.[83]

■   ■   ■

The NASA review board had conducted a thorough investigation, something critics didn't think was possible. And even after it was over, some in Congress were not too thrilled with the performance. "I am deeply disturbed by the report of the Apollo 204 Review Board," said Congressman Donald Rumsfeld of Illinois. "My concern is not that the board was unable to discover the precise cause of the accident during the course of its rigorous technical investigation, but, rather that the Board failed to examine, or at least report on, the fundamental conditions which permitted the accident to occur and which, moreover, resulted in the tragic deaths of three fine young men. The report's, and apparently the investigation's, shortcomings leave open the possibility of similar catastrophes in the future." On the Senate side, Walter Mondale accused NASA engineers of "criminal negligence," while J. William Fulbright called for a "full reappraisal of the space program."[84]

Itchy congressional trigger-fingers aside, the investigation lasted for four solid months, with the board totally immersed in its work. Investigators even set up an exact duplicate of the Apollo 1 capsule, with the exact same conditions, and sparked a fire to better figure out exactly what had happened. Stu Roosa, who saw the damage in Apollo 1 within an hour of the tragedy, noted that this particular test and the damage it caused "looked exactly like the real spacecraft. Everything was the same except the bodies." In April, the Apollo 204 Review Board released an official report that ran more than 2,300 pages and weighed 15 pounds, which laid out specific problems and made specific suggestions for how to fix them to make spaceflight safer in the future.[85]

North American Aviation, as might be expected, was not happy with the report. The company's president, Lee Atwood, actually ridiculed it. "I was staggered," he said. "The tone of the thing, and the implications and the imprecision, the inflammatory nature of it...[were] things that you'd expect out of the *National Inquirer.*" But Borman was very pleased with the board's effort. "Our report was an impartial analysis of the fire, and we listed a number of recommended changes in the design of Apollo's spacecraft. We didn't sweep a single mistake under the rug, and to this day I'm proud of the committee's honesty and integrity," he wrote in 1988. Buzz Aldrin agreed, noting that the members of the "extremely qualified" review board "were men of integrity" who did an "honest job." North American took a hit but so did NASA.[86]

The media certainly piled on when the report came out, which had the effect of defending NASA's own review process. "Taken literally," wrote the *New York Times*, "the dry technical prose of the report convicts those in charge of Project Apollo of incompetence and negligence." A big problem for the *Times* was the failure to classify the plugs-out test as hazardous. The *Boston Globe* criticized the "many deficiencies in design and engineering, manufacture and quality control. It had been feared that the...investigating board...would produce nothing but a whitewash. But its report could scarcely be more damning."[87]

Compounding the problem, Webb and top NASA officials had looked bad at the hearings and took their licks in the press. The *New York Times* lit into NASA for its conduct. Webb's contradictory testimony "strongly supports the belief that neither Congress nor the American people have been treated with full candor in NASA's reporting of what was going on in the space program. It may be recalled that the famous Phillips report became public property only as a result of a leak to Congressman William F. Ryan," the editors wrote. North American Aviation "was picked for the Apollo project by a few high officials—and not by a large group of technical experts as originally imagined." It was the result, they said, of influence peddling.[88]

And NASA's behavior and the media onslaught only hurt their status with Congress. Some members made a mocking acronym out of N.A.S.A.—"Never A Straight Answer." One New York congressman was especially upset. "We have had to take you at face value," said John Wydler. "Maybe it was our fault for taking you at face value." He was more direct with NASA's administrator. "I think you have a problem with secrecy, Mr. Webb." A representative from West Virginia advised the space agency that things would be different moving forward. "I intend to be much more skeptical of NASA in the future, on this program and others," he said. With this level of renewed skepticism, there was worry that NASA's budget would be in the crosshairs for the next fiscal year in 1968, as budgets for the Great Society and Vietnam were climbing. A number of astronauts would testify, including McDivitt, Schirra, and Slayton, which, it was hoped, would smooth things over. Yet none of them really enjoyed the experience. "It was exasperating having to hear a half-assed politician expound on the deficiencies of the Apollo program," said Schirra.[89]

But, out of all the astronauts who appeared, Borman became the star attraction. Webb needed someone to help pull the agency off the congressional cliff and Borman was the man he turned to. Before he testified to congressional committees in both houses of Congress, Webb gave him some advice. "Frank, the American people need to understand what happened just as we now understand. You are not to try in any way to hold back facts or color your testimony in NASA's favor. Just tell them exactly what your investigation group found, no holds barred, even if it makes NASA look bad." Borman may have faced his share of danger as a test pilot and astronaut but this was a whole new ballgame. "The congressional investigative hearings too often are premature, carelessly conducted, overplayed by news media catering to headline hunters, and motivated too much by sheer politics," he later wrote. Given this situation, he had a similar attitude as most of his colleagues. "I looked forward to testifying on Capitol Hill with all the eager anticipation of a man going to a dentist and facing certain tooth extraction."[90]

Despite his misgivings, Borman did not disappoint. At one point he asked the million-dollar question. "We are confident in our management, our engineering, and ourselves," he said to the nation's lawmakers. "I think the question is, are you confident in us?" That was a serious attention-getter. But before he was finished with his testimony, a few days later, he hit them with a crushing rhetorical blow. "Let's stop the witch hunt and get on with it." The room burst into applause.[91]

Other astronauts felt exactly the same way. In May, Al Shepard spoke at a NASA banquet for the top people in the program, the astronauts, aerospace executives, and members of the media. Glaring at a few of the assembled press, Shepard stated, "The time for recrimination is over. We have digested enough historical evidence. There is much to be done. Morale is high. Vision is still clear. And I say, let's get on with the job."[92]

When all was said and done, the witch hunt would end, and the Apollo program would move forward. Congress had its say, as it always does. Aside from critics, NASA did have its defenders on the Hill, particularly in the House. In addition to those who had come immediately to Gus's defense when he was accused of causing the fire, the hearing's chairman, Olin Teague, defended the agency from abuse by its detractors, especially with regard to the Phillips Report. "It is my understanding this is nothing more than a group of notes," he said. "There really is no Phillips Report." Furthermore, Teague did not call Joe Shea as a witness. "It is the prerogative of the chairman to call witness," he said to a fellow member in his own defense when questioned about it. "If the gentleman doesn't get the information he wants from the witnesses that are before us today I don't think he can get them anywhere in the world." Borman thought Teague "was magnificent—sympathetic and understanding, yet also firm in his probing for the facts."[93]

For its part, the Senate, where the bombshell revelations came, produced its own report, published on January 30, 1968. "No single person bears all of the responsibility for the Apollo 204 accident," the report stated. "It happened because many people made the mistake of failing to recognize a hazardous situation." Though three brave astronauts perished, because of their deaths, "manned space flight will be safer for those who follow them. The names of Grissom, White, and Chaffee are recorded in history and the most fitting memorial the country can leave these men is the success of the Apollo program—the goal for which they gave their lives."[94]

Three senators also published "Additional Views" within the Senate Report, including Mondale, who was by far the most critical. Calling the Phillips Report "the most significant single document involved in the Apollo 204 investigation," Mondale took a few shots at NASA brass and the program itself. "The Phillips report represented the most far-reaching and fundamental official criticism ever made of a major NASA program. The biggest and most ambitious NASA program of all—man's flight to the moon—was in deep and perilous trouble, and Congress was unaware of that fact," he wrote. Failing to notify Congress of the problems was a "unquestionably serious dereliction," raising "the question of whether the committee and the Congress are to be limited to only that information which NASA sees fit to provide or whether Congress will be supplied with complete and candid information regarding the basic problems and difficulties being experienced in various NASA programs." Making matters worse, when the Phillips Report became known, both "NASA and NAA officials attempted to mislead members of the committee." This effort to mislead, and a general "patronizing attitude toward Congress," will only hurt the effort by producing a loss of public and congressional confidence. Despite Mondale's rantings, in the end Congress didn't punish NASA but increased the space budget to help cover the cost of redesigning the spacecraft and adding a number of safety features, to the tune of half a billion dollars.[95]

■   ■   ■

With the investigations complete, and the end of all the flying accusations, NASA still had much work to do to get the program back on track and get to the moon before the end of the decade. The spacecraft had to be redesigned and changes had to be made at both the space agency and with North American Aviation. After NASA removed Joe Shea as head of the Apollo Spacecraft Program in Houston, the agency replaced him with the deputy director of the Manned Spacecraft Center in Houston, George Low, a brilliant engineer who had been one of the earliest voices pushing for a lunar landing flight. As Tom Stafford noted, "Low's style was completely different from Joe Shea's; for example, he communicated better with Robert Gilruth and other center directors, including them in a spacecraft change control board. He would also listen to input from the flight crews." Alan Shepard also thought highly of Low. "What George Low did was instill a sense of dedication and purpose among those working under and with him," he said. It would turn out to be one of NASA's smartest moves.[96]

As for the spacecraft, Borman would be asked to serve on the newly formed spacecraft redefinition team that would be based in Downey, California, at the North American Aviation plant. Borman accepted immediately. His main job was to make sure North American implemented the vast changes NASA demanded within the spacecraft. When he arrived at Downey, he found major issues. "North American was positively schizophrenic, populated by conscientious men who knew what they were doing and at least an equal number who didn't know their butts from third base." Borman also learned that many of the plant workers were taking their lunch break at a bar across the street and having beers with their meal. "I couldn't believe it. These guys were working on highly sensitive and complex equipment, requiring absolute perfection, and here they were guzzling beer right in the middle of their workday." Tom Stafford had a similar experience. "My first exposure to North American's Downey team made me think that some were more interested in what they were going to do off the job than on it." Borman met with the management of the workers' union and the midday drinking stopped. As did a lot of other problems with North American. "Borman's direct, often abrasive approach to problem solving undoubtedly let them know that the previous cozy relationship between Houston and Downey was over," wrote Buzz Aldrin.[97]

In the end, North American did not fight NASA and took their share of the blame for the tragedy. But many of their employees didn't like it one bit. They felt the blame lay with the space agency. "Many of us were disheartened," said Lawrence Korb, an engineer. "I wanted to write an article for *Life* magazine telling them all the things NASA did wrong, but I didn't. I recognized there is a time to bite your lip and take the blame. I thought that if the country thinks NASA is careless or incompetent, they might cancel the Apollo Program, but if they think NAA was sloppy, NASA could either shape up NAA or put a contractor over us."[98]

North American also made some major changes. The company changed its hiring practices, at NASA's request, and merged with Rockwell International to became North American Rockwell. They brought in a new boss who began to clean house. The program director, Harrison "Stormy" Storms, was removed from the head of the Apollo project and replaced with Bill Bergen from Martin. They also agreed to a reduction of $10 million in fees.[99]

The review board had made recommendations for changes and improvements in the new spacecraft design, and together, North American and NASA fixed all the problems with the spacecraft, implementing 1,341 design changes in the Block II. "We shouldn't have been flying the Block I," wrote Slayton, so they were all grounded. Jerry Goodman, a NASA engineer, worked on the spacecraft redefinition team at Downey. He, like the rest of the team, spent almost all of 1967 in California fixing the Block II spacecraft. "1967 was horrendous. I was hardly ever home," he said. But what transpired with the redesign, Goodman noted, was simply amazing. In a relatively short period of eighteen months, from May 1967, when the work began, to October 1968, when the first manned Apollo flight lifted off from the Cape, the entire capsule was redesigned. "It was a significant effort to fix all the problems," Goodman said. "I can't stress that enough. It was miraculous to do what we did in that span of time."[100]

The newly redesigned spacecraft might have looked similar to the AS-204, but there was a world of difference beneath the skin. All the wiring was improved and re-routed with better insulation and shielding inside metal troughs. Most flammable materials, like Velcro, were either greatly limited or eliminated altogether, including the astronaut spacesuits, replaced with non-flammable materials like Beta cloth. Velcro was limited to no more than two-inch by two-inch squares that had to be spaced apart from one another. Nothing larger than that. The explosive 100 percent oxygen in the capsule was replaced with a two-gas system, while the astronauts breathed 100 oxygen in their suits. The nitrogen, which was detrimental to astronauts in space, was filtered out while en route to orbit. There was also an emergency oxygen system added to the spacecraft and an emergency venting system so that cabin pressure could be reduced quickly. The capsule also got a new hatch that could be opened in three seconds, which added 1500 pounds of weight to the spacecraft but was a necessary trade-off. In all, NASA spent half a billion dollars on the redesign and safety features in the spacecraft and elsewhere.[101]

"We spent a great deal of time to find the source of the fire," said George Mueller, "but it was literally a time bomb waiting to go off.... Unfortunately, it had to happen then. But we did persevere, and I would say that the good thing that came out of it was we really understood what causes fires on spacecraft. We redid most of the wiring, not that we knew the wiring was at fault, but rather we redid the wiring on Apollo and did it more professionally than the first time around. I think that's probably why the Apollo Program was relatively accident free" after the fire.[102]

NASA also made three other improvements. There was an increase in fire and medical personnel able to respond to an emergency more quickly. An escape system was added to allow the astronauts to get off the tower in case of fire or explosion. Until Apollo 1, there had only been one way down—the same elevator that took astronauts to the top. After the tragedy, a newly installed gondola could sling the crew four hundred feet away from the rocket in about six seconds. It remained in place through the space shuttle program and will be used for all future space launch systems.[103]

The final change was the important personnel re-hire of Guenter Wendt, a German engineer who worked for McDonnell and had served as the pad leader for all manned launches in Mercury and Gemini, since McDonnell built both spacecraft. He was a tough, abrasive guy who did not tolerate foolishness. The astronauts referred to him as the "launch pad fuhrer." Astronaut Pete Conrad described him best, "It's easy to get along with Guenter—all you have to do is agree with him." But he was close to the astronauts. Wendt was the last person the astronauts saw before the hatch was closed. "The fire hit me very hard because when you know the individuals, you know you have horsed around with them, they are your friends," he said.[104]

For Apollo, Wendt had not been the pad leader because the capsule was built by North American Aviation, which had its own pad leader. As Harry Hurt has written, "Many believe that the Apollo 1 fire, which occurred shortly after North American took over as prime contractor, would never have been allowed to happen if Wendt had been in command." Yet he would never say that the fire would have been prevented if he had been there. There was simply no way he could know that. But he did know about the dangers of pure oxygen in the spacecraft. He even warned technicians about the hazards, telling them not to smoke for at least four hours after a test "because the oxygen saturates your clothes and polyester clothing goes up in flames." After the fire, when NASA made sweeping changes, Wendt was re-hired and served as pad leader for all manned Apollo missions.[105]

Not every astronaut was giddy about the salutary changes that came out of the fire, though. "For those of us who knew Gus, Ed, and Roger," said Young, these changes "gave us small comfort. A long string of mistakes and misthinkings caused the accident to happen. No one person or organization took the whole blame; there was enough blame to go around." Slayton tried his best to keep everyone's spirits up. After the report was released, he called a meeting attended by the top eighteen astronauts. "Gentlemen, we won't make the same mistake twice," Slayton announced. "All of us wanted to believe him," Young said, "but we knew, deep down, that mistakes would continue to happen." They just had to work hard to try to prevent as many big ones as they could. "That was what engineering was all about: preventing the mistakes that led to big problems and major failures."[106]

The astronauts exhibited different reactions to the fire and the recovery. Some were positive; others less so. And at least one astronaut was downright hostile. Donn Eisele, who might have ended up in the spacecraft during the fire had it not been for a shoulder injury, made a number of nasty comments about North American in his autobiography, *Apollo Pilot*, which he did not complete before his premature death in 1987 at the age of 57. Perhaps experiencing "survivor's guilt," or exhibiting anger with NASA over the fact that he was never allowed to fly again after his lone mission on Apollo 7, Eisele lowered the boom on everyone, particularly on George Mueller and Joe Shea. "The real tragedy of the 204 fire was that it was so preventable, so unnecessary—almost criminal, in fact. The incredibly incompetent management, which, in my opinion, led to the fire, arose out of George Mueller's naïve stupidity and Joe Shea's colossal ego," he wrote. "Mueller would dictate some impossible schedules for spacecraft delivery, checkout, and launch. Shea was vain enough to think he could meet them. And the contractors were dishonest or equally stupid, or both, for not telling NASA they couldn't hack it."[107]

"It's ironic" he said, "that the first major change made to the spacecraft after the Apollo 1 fire was the simple, one-piece hatch we had asked for two years earlier. And you should have seen the North American weenies. They crowed about the 'new' hatch until one would think they had invented the wheel."[108] But Shea hadn't been interested in hearing astronaut feedback. "He seemed to have a peculiarly contemptuous suspicion of astronauts, anyway. How could dumb-ass pilots know anything? Most of them didn't even have Ph.D.s!" Things would only get better, according to Eisele, once Shea became the scapegoat and got canned.

Although North American Aviation, the company that built the spacecraft, should shoulder the majority of the blame for the accident, as should any manufacturer that produces a product that kills its users, NASA, too, was culpable, though not as directly. North American was certainly guilty of extremely sloppy and at times careless work, and NASA can be cited for allowing politics to cloud good judgment and for pushing too hard, too fast. There was plenty of time to make Kennedy's deadline at the end of 1969. There was no need to push that hard, that fast. NASA was also guilty of misleading Congress on a number of issues, some understandable, others not. And, most importantly, the space agency was guilty of failing to heed the warnings about the dangers of fire in the spacecraft with 100 percent oxygen at higher pressures. That is beyond dispute.

In the end, Betty Grissom placed a lot of blame on North American Aviation for killing her husband and filed a $10 million lawsuit against the company. The suit was settled for $350,000, which was also paid to the families of Ed White and Roger Chaffee. Betty also held a major grudge against NASA. For the rest of her life, she never again set foot on NASA property.[109]

Despite the tragedy, the struggles with Congress, and the politics behind the North American deal, NASA was poised to land on the moon before the end of the decade. The fire "threatened for a time to replace our excitement with national despondency," wrote Lyndon Johnson, "but this did not happen." The country eventually rallied behind the space program to meet Kennedy's goal.[110]

His own congressional troubles notwithstanding, Jim Webb deserves a share of the credit for getting America to the moon, before and after the fire. Without his administrative skills, there can be little doubt that attempting to meet Kennedy's deadline would likely have ended in failure.

After the tragedy, he was able to salvage congressional, and public, support for the program, maintain adequate levels of spending, and keep the agency on track toward its ultimate goal. And although he left the agency in 1968, the year before the lunar landing, he was as instrumental as anyone in making it a success. He steered the ship through very treacherous waters in 1967 and put NASA in position to land on the moon in 1969. That being said, later revelations would tarnish his star in the eyes of many.

THE POLITICS:
A WEBB OF INTRIGUE

Bobby Baker was never elected to the United States Senate, or to any office for that matter, yet he was one of the most powerful individuals within that hallowed institution. A close confidante and protégé of then-Senator Lyndon Johnson beginning in 1948, Baker became so close to LBJ that he earned the nickname "Little Lyndon." His power and influence in the upper chamber were so great, in fact, that he was once dubbed the "101st Senator." He knew how to get things done, one way or another.

He came from the "South Carolina backwaters," he wrote in his autobiography, "and within a relatively short period gained the reputation of being one of the most powerful men in government." Yet in his telling, there "wasn't any magic in it." Just industry, hard work, ambition, and the fact that he really cared about his job. Others would have a far different take on his rise to power. Beginning his career as a Senate page when he was fourteen years old, Baker moved up the totem pole over the next few years, serving in various Senate jobs: messenger to the Senate minority, assistant to the doorkeeper, chief telephone page to the majority, and assistant clerk to the majority conference. Before long, he knew more about the Senate's "day-to-day operations than many of its members," he said. Then he met Lyndon Johnson in December 1948 when he was but twenty years old and the Senator-elect from Texas was forty. His career would skyrocket as a result, though he ultimately flew too close to the sun.[1]

Soon after his legendary election to the Senate in November 1948 but before his January swearing-in, then Congressman Johnson was ambitious to climb the rungs of the hierarchy of his new position from the get-go. He set out to find "who's the power" in the chamber. He learned very quickly that Baker was the young man he needed to know. "Mr. Baker, I understand you know where the bodies are buried in the Senate," he said on the telephone that December day, the first words spoken by Johnson to Baker. "I'd appreciate it if you'd come to my office and talk with me." Baker did, sitting in Johnson's House office for a full two hours of conversation, mostly initiated by LBJ. "I wanted to meet you. My spies tell me you're the smartest son of a bitch over there," Johnson told him when they first met. "I said, 'Well, whoever told you that lied,'" Baker responded. But there was a lot he did know—and most importantly, he knew the kind of information Johnson needed. "I said, 'I know all of the staff on our side. I know who the drunks are. And I know whose word is good.' He said, 'You're the man I want to know.' So we became great friends."[2]

Flattery aside, Baker was a man of influence, even at such a young age, and would serve by Johnson's side as the secretary for the Senate majority for more than a decade. He became Johnson's "strong right arm," the "first man I see in the morning and the last man I see at night," LBJ admitted while he was in the Senate. They were so close that Baker lived next door to Johnson and named two of his kids Lyndon and Lynda. Yet Baker was also "a young man of breathtakingly few scruples," wrote author Joan Mellen. And the young man on the rise with few moral qualms would find himself neck-deep in the political intrigue that lay behind the Apollo 1 fire and landed in the clink.[3]

■   ■   ■

Conspicuously absent in all the Apollo 1 investigations was the politics swirling around the selection of North American Aviation as the prime contractor for the spacecraft. There were journalists who looked into it at the time, most notably Pulitzer Prize–winner Clark R. Mollenhoff, who believed politics was behind the deal, but nothing in any official capacity, and for obvious reasons. As the rumors began to seep out, and scandals began to emerge, several Republican members of Congress pressed for an investigation into how exactly North American had received the contract, but their pleas went largely unanswered. The politics, though, were perhaps the key to the tragedy and would pull back the curtain on what was going on behind the scenes.

Norman Mailer thought so. Commenting on the aftermath of the tragedy he wrote, "Of course, one did not have to look for a psychology of machines. There was a psychology of politicians as well, there was that complacency which finds its way into the most serious tests of the most critical materials when a large corporation has the highest influence in the highest places, as indeed which large corporation does not. One could look for the villain everywhere, even in the White House. There were more than a few to whisper maliciously that the bag if ever opened would have blasted Lyndon Johnson to the moon. And Bobby Baker could have been his launch vehicle."[4]

■   ■   ■

As we have seen, NASA awarded the contract to build the Apollo command and service modules, as well as the Saturn V second stage, to North American Aviation in Downey, California, in the fall of 1961. The previous July, Gilruth's Space Task Group sent out an RFP, Request for Proposals, to fourteen aerospace companies and five submitted bids to build the Apollo spacecraft. The five were McDonnell, which built the

Mercury and Gemini spacecraft; General Electric; General Dynamics; Martin Marietta of Baltimore; and North American Aviation. The deadline for submitting a bid was October 9. Robert Seamans approved an eleven-member Source Evaluation Board headed by Max Faget that consisted of 190 experts to evaluate and rate each bid. And it would take some time before the process was completed. "Any given contractor might submit a stack of reports and papers two to three feet high," Seamans said. And the board had to go through every bit of it. Each company also had to present its proposal before the board and answer tough questions from the likes of Gilruth and Faget, inquiries like, "What single problem do your people identify as *the* most difficult task in getting man to the moon?" or how did each company propose to deal with micrometeorites that could destroy the spacecraft and kill the astronauts?[5]

When the board finished its work and weighed everything in the balance, Martin came in first, with a rating of 6.9. North American finished second, tied with General Dynamics at 6.6. General Electric and McDonnell finished up the list tied at 6.4. Therefore, the board recommended Martin. The NASA team further recommended North American Aviation as the second choice, if Martin were not awarded for whatever reason or failed to deliver under the terms of the contract. Martin, though, had been working on getting the spacecraft contract for two years, spending millions researching various concepts for the capsule and service module. Word soon leaked out that they had received the highest score from NASA when a Maryland congressman called the head of the company to offer his hearty congratulations. Naturally employees were overjoyed that they would be getting the massive contract worth billions. They were on the highest peak; the next day they would be in the lowest valley.[6]

Unbeknownst to Martin's employees, things were happening behind the scenes to cut the company out of the contract. Bob Piland, the deputy project manager of the Apollo Spacecraft Program, called Tom Markley, who served as a secretary for one of the evaluation board's subcommittees and asked that the scores be re-evaluated. What NASA hierarchy wanted was for more weight to be given to a firm's experience producing experimental aircraft, like the X-15 built by North American. This would certainly boost North American Aviation's score, which is exactly what motivated the reevaluation.[7]

Harrison "Stormy" Storms, the head of North American's space division, was at the White House attending a ceremony for legendary aviator Scott Crossfield, who was the chief test pilot for the company. President Kennedy was presenting Crossfield with the Harmon Trophy for his feats piloting North American's X-15, which flew to the edge of space. Storms was in a depressed mood because he had heard the talk that the contract had gone to Martin. Bob Seamans was also at the ceremony, as a representative of NASA, and he asked Storms to come by his office later that day. Storms suspected that Seamans wanted to tell him that North American had not gotten the big contract. But later that afternoon, sitting in his office at NASA headquarters, Seamans broke the news to him. "I just wanted to tell you personally while you were here in town. You've won Apollo." Storms was stunned. It was the last thing he expected to hear.[8]

On November 28, 1961, NASA publicly announced that the contract would go to North American Aviation, and the web of intrigue surrounding the deal was more than a little unnerving, to say the least. And it certainly didn't help matters, at least from a public-relations standpoint, that, on the night the contract was awarded, North American employees were seen with baseball caps embroidered with a unique NASA logo: NA$A. This type of behavior is certainly what Joe Shea meant when he said North American had dollar signs in their eyes. With contract in hand, North American would pull in $110 million per month, sixty percent of its total income, on NASA contracts. Perceptions would only get worse. Aside from the score re-evaluation, it seemed that North American had had the inside track on the contract before the process ever started. During the selection progression, one of the review panels consisted of an engineer from the Space Task Group and astronaut Alan Shepard. During the meeting Shepard lost his patience with all the back-and-forth. "This is all a waste of time," he said. "It doesn't make any difference what the score is. North American is going to win."[9]

And that wasn't the only evidence that North American might have had it in the bag all along. As Charles Murray and Catherine Cox noted,

> Another man who had been with the Space Task Group for about two weeks remembered walking to lunch during the oral presentations, before North American had even given its pitch, and hearing another astronaut, Gus Grissom, say to his companions, "By God, I'm going to do everything I can do to make sure North American gets this contract." The man was shocked. What kind of outfit am I in, anyway? he asked himself.
>
> Others recall the ways in which, in those years, North American seemed to act as if it had the government in its pocket. One senior NASA official remembered being present in [NASA engineer] Brainerd Holmes's office when Holmes told a senior executive from North American that he would not be permitted to bid on the lunar module because of North American's backlog of work. The executive is supposed to have said, explicitly, "If you don't let me bid, I'll have your job"—which seemed to the observer to be a strange thing for a corporation executive to think he could say to an official of the government.[10]

Murray and Cox, though, shy away from any taint of corruption and politics in the awarding of the command and service module (CSM) contract to North American. "No one ever proved that anything shady had gone on, but the coincidences were noted and the rumors persisted," they wrote.

But when the full inside story is laid out, rather than speculation as to what might have transpired, it certainly seems that some backroom dealing had taken place. For one thing, not every astronaut believed things were on the level, at least as far as the spacecraft contract was concerned. Several thought NASA's move made little sense. Some were even told by NASA engineers that they thought the Martin Company had a better design. It smacked of politics, pure and simple. "Decision-making at the upper levels of government is beyond me," astronaut Wally Schirra said. "I couldn't fathom the politics." He suspected that "North American got the contract because of politics," because "California companies hadn't got a big bite of the space program" at that point. For John Young, who would fly six spaceflight missions before his retirement, there was "a lot of politics in big expensive government contract awards" given by NASA, or any other part of government for that matter. The whole episode reeked of political favoritism and influence peddling, if not outright corruption, and it was worse than he, Schirra, or anyone else ever realized.[11]

The major players in the whole deal were all friends and intimates. Norman Mailer called it "the purest kind of guilt by association. But what association!" The group included Vice President Lyndon Johnson, NASA chief Jim Webb, Senator Robert Kerr of Oklahoma, North American's chief lobbyist Fred Black, and, perhaps most importantly, Bobby Baker. In fact, the New York Times later pointed out that "the shadow of Bobby Baker...darkened the history of North American's relations with NASA." He was the center of a circle of corruption and shady dealings. And, making matters worse, a number of these intimates met with LBJ in his vice-presidential office in an official, on-the-record capacity. Black, Baker, and Storms met with Johnson on August 21, 1963. The previous month, Lee Atwood and Baker met with the vice president. The same day, LBJ met with Baker three more times alone. It is quite possible that these meetings concerned the vast amounts of corruption that was beginning to spill out in the press and would eventually lead to criminal charges against Black and Baker.[12]

For his part, Johnson's fingerprints might not have been on the actual deed but his presence was certainly felt. He was friends with Kerr, who had succeeded him as chairman of the Space Committee in the Senate, and had essentially picked Webb, a Kerr-McGee executive, for NASA. In fact, according to Baker, in a Senate oral history recorded in 2009 and 2010, when LBJ asked Kerr about a NASA administrator, Kerr replied, "Lyndon, I've got the man." That man was Jim Webb.[13]

Black, as North American's top lobbyist, had many friends on Capitol Hill. He was close with Congressman Gerald Ford, the future president, and Dewey Short, who had been chairman of the Armed Services Committee and assistant secretary of the Army. Black was also a close friend and neighbor of Lyndon Johnson for years in the same neighborhood in northwest Washington, D.C. When Johnson became vice president, in the days before the nation's second office had an official residence, he sold his home and bought a new one, a mansion known as The Elms, which would be much more lavish for cocktail parties and other social events. Within months, Black sold his home and moved once again to a house next to Johnson. Their property adjoined and shared a backyard fence. Bobby Baker did likewise. He lived one street over. The band stayed together.[14]

Senator Kerr, who Baker said "knew more ways to skin the legislative cat than most," was the biggest fat cat behind the whole deal. He was rich, the most powerful man in Oklahoma, and supremely arrogant. He had served as governor of Oklahoma from 1943 to 1947, then was elected to the U.S. Senate in 1948, the same year as Lyndon Johnson. In 1952, he tried to grab the Democratic nomination for president, but garnered no more than sixty-nine ballots at the convention and quickly withdrew. "Nothing on earth made him do that but his ego," one of Kerr's friends told the *New York Times*. "Bob Kerr had as big an ego as any man you'll find." Failing in the presidential sweepstakes, he worked to gain all the power he could in the Senate, which was considerably easier after LBJ left in 1961, and he succeeded to the point that some called him the "uncrowned king of the Senate," an epitaph he loved. "I represent myself first," he once said, "the state of Oklahoma second, and the people of the United States third, and don't you forget it." He never minded putting someone in their place, and he had ample ways of doing so, for in addition to chairing the Space Committee, Kerr also sat on other powerful committees, like Finance and Public Works.[15]

With vast amounts of power and influence, Senator Kerr decided he wanted the Apollo spacecraft contract to go to North American and was hard at work on the deal well before it was awarded, probably because North American was willing to give him everything he wanted. The reason had already been made clear to them a year or so earlier. North American executives came to Washington to speak to Kerr about saving one of their planes from budget cuts, the B-70. After making their case as to why the bomber should be saved, Kerr reminded them what was really important. "Gentlemen," he said, "you haven't told me what's in this for Oklahoma," and, by extension, himself.[16]

The Senator's idea was for North American, in return for the contract, to build a complex of plants in Oklahoma that would hire twenty thousand Oklahomans to construct Apollo components. According to Baker, "Kerr then pressured James Webb, director of the National Aeronautics and Space Administration—and, conveniently, an executive on loan to the government from the Senator's own Kerr-McGee Company—to give North American Aviation favorable treatment on the Apollo contract. Webb certainly did nothing to stand in the way, and North American got it." As was his policy, Webb later denied any connection with Black and the awarding of the contract. "Black was sent to me by Senator Kerr on three different occasions. We turned every one of them down." Yet North American still came away with the prized contract.[17]

New York Times reporter John Noble Wilford, like Murray and Cox, believed there was "no evidence to suggest Baker and Black had exerted any illegal influence" on Webb or anyone else in order to gain the contract for North American. To bolster his case, Wilford asked Webb if there were any "possible under-the-table deals involving Apollo." Webb's unsurprising response was again to deny any wrongdoing. "There were absolutely no political implications in any of our contract decisions," he told Wilford. How could anyone expect him to say anything any different?[18]

At other times, Webb was more defensive about such inquiries. "A lot of senators and congressmen talked to me about the Apollo contract. I was willing to talk to everyone about it," he said. "There are a lot of people who have tried to give the impression that I was not independent, and that I was in Senator Kerr's pocket because I worked for the Kerr-McGee Enterprises." It certainly looked like it. Not only did Webb do "nothing to stand in the way," he tried to cover for the deal in front of Congress. Clark Mollenhoff wrote that Webb was "irritable and defensive under heavy criticism" during his congressional testimony. He "tended to defend North American Aviation and has declared that it is more important 'to get on with the job'" than it was "to go back and pinpoint responsibility for all negligence." When asked by Senator Margaret Chase Smith on April 17 if North American was the first choice of the evaluation board, Webb was less than truthful, replying, "Yes. It was the recommended company," likely figuring that no one would ever find out that it wasn't, at least not initially. But Senator Smith later discovered the truth and called Webb back to testify three weeks later. At that time, the NASA administrator had to retract his previous statement, admitting that the evaluation board had chosen Martin, not North American, and that he had overruled them, citing the company's "know-how." Webb, in a move to toss his subordinates under the bus, asked for the reevaluation at the behest of Gilruth, Low, Williams, Dr. Hugh Dryden, and others. But was this the only rationale? If so, why would he not simply tell Congress the truth rather than attempt to hide it? Because the truth looked bad and he knew it.[19]

This is not to suggest that Jim Webb, the very able administrator of NASA, was himself neck-deep in corruption and underhanded deals. He certainly had to know what was going on, but he did not profit by any of it, nor has anyone made an allegation against him that he did, at least not a credible one. But, according to his deputy, Bob Seamans, Webb was "a consummate political animal." He wanted to make sure he was friendly with the members of Congress who were on the space and defense committees who oversaw NASA and determined budgetary requests, and he wanted to make sure that nothing NASA did would upset that relationship. As Mike Gray has written, "Like an exposed pawn on the front rank of the chess game, he was sensitive to the alignment of power, and he knew that the fateful decisions had to resonate correctly off the foundation of power." Which is why he didn't fight the location of the Manned Spacecraft Center in Houston but understood the delicate politics involved, nor did he make any fuss over the contract for North American. He knew how the game was played.[20]

With Webb's tacit support, Kerr was willing to do everything in his power to secure his new space enterprise for Oklahoma, vowing that nothing would stand in his way, not even geography. When he met with North American executives to formulate the plan for the plants and jobs before the contract was awarded, one of the company's managers agreed to his demands but had to admit there might be a major issue. "However, we've got a problem, Senator. Some of the equipment will be so large it can't be moved except by water. And Oklahoma is land-locked." Kerr never batted an eye, responding in words befitting an "uncrowned" monarch, he told them: "Tell me how wide and deep you want the ditch and the government will dig it."[21]

An impossible task for a mere mortal but not for Bob Kerr, who also happened to be the chairman of the Rivers and Harbors subcommittee of the Senate Public Works Committee. He was as good as his word, and today the ditch is known as the McClellan–Kerr Arkansas River Navigation System, a canal network that runs 445 miles with 17 locks connecting the Mississippi River to the Port of Catoosa in Tulsa. It cost the American taxpayer $1.2 billion by the time it was completed in 1971, even though the largest component of the Apollo project that North American would build was in California. No matter, for in February 1962, North American Aviation conveniently announced that it had chosen Tulsa "as the site of a major plant in a complex it was developing to build about $1 billion worth of equipment for the Apollo lunar project," purchasing 300 acres near the canal line. But the project also put some more cash into the pockets of the well-connected. According to the *New York Times*, "Senator Kerr and his associates, by coincidence, owned hundreds of acres of property along the line of the barge canal." Eleven-hundred acres to be exact. The fact did not become public knowledge until after Kerr died on January 1, 1963. When his estate was probated in court, it was discovered that the land was in the name of an attorney in Oklahoma City, a likely tactic to hide the Senator's involvement. This land would be purchased by the government to build the canal, and one would think at an exorbitant price. "Prior to that time," wrote Mollenhoff, "there was no knowledge that Kerr...had any personal financial interest in the Arkansas River project." Now everyone knew.[22]

But Kerr, Black, and Baker also had another deal in the works involving North American Aviation, which was tied directly to the Apollo contract. In late 1961, Baker, who was always "wheeling and dealing," which, incidentally, was the name of his memoir, entered into a "more profitable" investment than his previous enterprises. "The opportunity came about through Senator Kerr and Fred Black," he wrote. "Fred...drew a $300,000 salary from North American Aviation. He was paid another $75,000 or so per year by Melpar, Inc., a subsidiary of North American." Baker estimated that Black made a half a million dollars annually in the late 50s and early 60s. But, Baker recalled, that much "just wasn't enough money for Fred Black. He was a playboy of the first order; if he couldn't go first class, then he wouldn't take the trip." Many at North American, including "Stormy" Storms couldn't stand him. One engineer said Black "would have sold his grandmother if he could make a buck.[23]

The original plan was for Kerr and Black to go into the vending machine business together. Kerr knew it would be a lucrative enterprise because they could get all the business from companies that had defense contracts through his influence. But, after receiving some obvious legal advice, Kerr dropped his part of the deal. He then told Black, "I want to help Bobby Baker. I'll get you the financing if you guys want to go into the vending machine business. There's a fortune to be made." The new business venture would be called Serv-U Corporation. Kerr wanted to help make Baker a millionaire, mainly because Baker was the same help to him in the Senate as he had been for Lyndon Johnson. LBJ thought of Baker like a son. So did the Oklahoma Senator. "Kerr loved Baker. Bobby was his ears, nose, throat, hands, and feet in the Senate," Black said. But Baker was in trouble in some of his previous business investments, especially the Carousel Hotel in Ocean City, Maryland, where he was straining to make bank payments of $8,000 a month. Serv-U would help put a lot of cash in Bobby Baker's pocket, as well as Black's. And Baker knew how to get customers as well as Kerr. As one competitor later alleged, as secretary of the majority, Bobby Baker "was able to, and did, represent...that he was in a position to assist in securing defense contracts."[24]

With Serv-U on the drawing board, Black was smooth enough of an operator to use the Apollo contract as leverage to obtain the North American vending contracts. When told that "Stormy" Storms wanted to bid on the spacecraft contract, Black called Lee Atwood, the president of the company, and told him that he was hearing from his sources that the odds were long for North American, which was unlikely, but, he added, their star would rise if they contracted with Serv-U for vending machines in all their facilities. Black just didn't bother to tell them that he was a co-owner with Bobby Baker.[25]

North American, which already had vending machine services in all of its facilities supplied by another company, severed those ties and awarded a contract to Serv-U on December 1, 1961, just three days after receiving the Apollo contract from NASA. In a 1965 phone conversation with LBJ aide Mildred Stegall, Baker admitted that it came about through the efforts of both Kerr and Webb. With neither a single employee nor any vending machines to install anywhere, Serv-U still received the contract.[26]

To fix the problem and get Serv-U up and running, Kerr knew just what to do. He arranged for a large loan for Black and Baker from Fidelity Loan and Trust in Oklahoma City, a bank in which he owned stock. In fact, Kerr-McGee had invested $1.6 million in the bank under Webb's direction when he was an executive at the company before his days at NASA. Therefore, Kerr treated it like his own personal piggy bank. In the spring of 1962, Black and Baker got a loan for $175,000 to purchase the vending machines. They later received another loan of $275,000 to expand the company's operations. And, coincidentally, one of the top twenty stockholders in Fidelity was Jim Webb, who still owned 1,866 shares in 1964 and had once been on the board of directors before he was at NASA. In a 1962 report to the House Banking Committee, Webb held a paper value of $790,000 in bank stock, which he never relinquished when he took over the role as head of the nation's space agency.[27]

Before long, Serv-U Corporation took on partners, a couple of whom were casino owners in Las Vegas and probably had serious mob ties, and soon it was making $3 million per year. Baker himself owned 28.5 percent of the stock. However, a lawsuit by a rival company revealed the partnership, which led to a Senate investigation and the eventual indictment of both Black and Baker for income tax evasion. When the dam broke, North American dumped the Serv-U contract and fired Black, who beat the tax rap, only to be convicted in 1982 of attempting to launder $1 million in Colombian cocaine money. He served seven years in prison. Baker wasn't quite as lucky. He was convicted of the tax charge on January 29, 1967, two days after the fire, and spent a year and a half in prison. And yet as much as he had done for Lyndon Johnson, Baker was on his own during his legal ordeal. "He acted like he didn't even know me," Baker later said of his former close friend.[28]

The emergence of what came to be known as the Bobby Baker Scandal blew the lid off many of the backroom deals and had some powerful and influential men running for cover, including Jim Webb. As he had done before Congress and to the press, Webb consistently denied ever having any dealings with anything remotely unethical, such as the reevaluation and awarding of the Apollo contract. He also denied having anything to do with Bobby Baker or knowing about any such deal with Serv-U, even though it was reported that he approved the deal and that he also owned stock in the same Fidelity Bank that financed the venture. He only learned about it, he later admitted, when he "read about it in the papers." As things began to unravel in 1967, he stated that he wasn't aware "until recently" that Fidelity was a major depositor of North American funds, as well as other space-related contractors who did business in Oklahoma.[29]

Webb also tried to downplay his role as an executive at Kerr-McGee before he was NASA's administrator, seemingly in an attempt to distance himself from Kerr. He explained that he was only working part-time before being tapped by JFK. "I was not in any way serving in an executive capacity in the period just before I joined NASA," he said. In addition to being a part-timer, he claimed that he was so busy with other things that he was not paying attention to what was going on with Fidelity and was not aware of the dealings with Black and Baker. But much of this talk would not wash.[30]

Despite Webb's pleadings of ignorance, Mollenhoff was not about to let him off the hook, writing that "the events of the last few weeks of the investigation of the Apollo tragedy have resulted in a lack of faith in his assurances" that the process had not been political in any way.

The FBI was well aware of what was transpiring and had been keeping close tabs on Fred Black, including with the use of wiretaps. After bugging one of his hotel rooms, agents heard a conversation between Black and McGee in which Black grumbled about the fact that Webb's attitude had changed toward him since Kerr passed away. "Since the old man died, this fellow Webb has gotten weaker and weaker where the state of Oklahoma is concerned," he said on the tape. Even though Kerr was gone, Black wanted more out of the association between himself and NASA. He had sent some additional proposals to Webb for approval but after Kerr died only about a third of what he wanted had been approved.

Then Webb quit dealing with him. "I'm getting concerned about a few things in Oklahoma City itself," Black said. "NASA is not helping us. When the senator was alive, he'd be helping," an obvious reference to Webb. But he was gone, and Webb saw no need to continue feeding Black or Baker. But the previous cozy relationship seems to prove that Webb was involved in helping Kerr get everything he wanted. He might not have been dirty like the other players, but Webb, at the very least, as Baker said, did nothing to stand in the way.[31]

Could this level of underhanded political dealings have played a major part in the disaster? Had the contract gone to Martin, as it likely would have, might the tragic outcome have been different? John Young certainly thought so. "I can't help but wonder what would have happened if the awarding of the original prime contract for the Apollo command and service module had gone differently," he wrote later. "Maybe the problems with the command module would just have been different ones, of a kind that could also have led to disaster and death. We'll never know. But I'll always wonder."[32]

Sadly, Young's thoughts were right. This was not a classic case of a simple government boondoggle whereby a project cost four or five times its original estimate, with no harm done to anyone but American taxpayers. No, this was, as stated by the Washington *Sunday Star*, a case where "know-who" won out over "know-how," an unscrupulous deal that cost three American astronauts their lives and nearly killed the U.S. space program. It is a sad state of affairs when something as noble and majestic as a national effort to get to the moon is tainted by the ugly game of politics. But it was, and it killed Gus, Ed, and Roger.

THE MOON:
"THE *EAGLE* HAS LANDED"

HERO. The word itself conjures up images of greatness. Perhaps the act of an unselfish martyr on a distant battlefield, or the lifesaving effort by first responders to save a life here at home. Some think of sports and the spectacular feats we see on the athletic field, while others look at far more selfless sacrifices.

But what does it truly mean to be a hero? *Webster's Dictionary* defines the word as "a person admired for achievements and noble qualities" and "one who shows great courage." A rather vague definition that could fit almost any situation, which is reflective of our modern world which has a great tendency to cheapen the word and extend it to persons not really deserving of it.

In the land of our old Cold War foes, the word was taken much more seriously and was used to signify the nation's highest honor—"Hero of the Soviet Union." And only those truly deserving of it would ever carry the distinction.

Robert Seamans, deputy administrator for NASA at the time of the fire, defined it this way in his autobiography: "Heroes, to me, are people who have tried to beat the odds, people who have made the most of what they have."[1]

But one of the best definitions comes from Pastor John MacArthur. "Real heroes are people whose efforts and sacrifices save lives, alter destinies, change history, or shift the course of history for the better." To change history or to shift its course for the betterment of our country. All of America's astronauts fully embodied these concepts of heroism. And every American should agree that Gus Grissom, Ed White, and Roger Chaffee qualify for the title.[2]

As *Life* magazine wrote soon after the fire, "Grissom...White... Chaffee...They bought the farm right on the pad, cooked in the silvery furnaces of their spacesuits, killed in a practice run before they could ever know the surge of their great Apollo craft driving upward to orbit. But put these astronauts high on the list of the men who really count."[3]

Anyone willing to climb aboard a largely untested spacecraft, sitting atop a rocket filled with millions of pounds of highly combustible and explosive fuel that could detonate at any moment during the flight, to be flung at the highest speed man can achieve into the dangerous void of space, then plunge back into earth's atmosphere protected by a heat shield where one is mere inches from outside temperatures reaching thousands of degrees, hoping that parachutes open at the proper time to gently guide the capsule down to the water for a soft landing, all for the betterment of our country, certainly counts as an American hero.

And true heroes are what the country needed in the late 1960s. While astronauts were seemingly doing the impossible for NASA, the country was under siege. The main antagonist was the war in Vietnam. In 1967, the United States had more than 485,000 troops in Southeast Asia, an increase of 100,000 since 1966. The troop level would hit its peak in 1968 with more than 536,000. During 1966, more than 6,000 American personnel lost their lives. More than 11,000 would die in 1967, and 16,500 in 1968. And, with the Tet Offensive, coming almost exactly one year after the Apollo 1 fire, there seemed to be no end in sight. When all was said and done, the moon race cost America roughly $25.5 billion, about $150 billion in 2020 dollars. The price tag for Vietnam was $138 billion, roughly $1 trillion today. At home, American cities were engulfed in the fires of race riots and civil unrest over the war and racial inequality. The Watts section of Los Angeles in 1965 suffered mass destruction, with 1,000 injures and 34 deaths. Riots in Detroit in 1966 saw 43 killed. In 1968, violence broke out in 120 cities across the country after the murder of Dr. Martin Luther King Jr.

But America's space program forged ahead and by 1968 provided the nation with the only good news in what seemed like an eternity. In October 1968, twenty-one months after the fire, NASA was back in space with Apollo 7, led by the Apollo 1 backup crew of Wally Schirra, Donn Eisele, and Walter Cunningham. Their mission, lasting nearly eleven days, was the one Gus, Ed, and Roger would have flown. And under Schirra's stern hand, the flight was nearly flawless.

After the success of Apollo 7, NASA took its boldest step, sending Apollo 8 to the moon in December 1968 under the command of Frank Borman, along with Jim Lovell and Bill Anders. Boosted into space by the mighty Saturn V, Apollo 8 circled the moon ten times and returned to earth safely, the crew becoming the first humans to venture into deep space. They were also the first to see the far side of the moon.

In March 1969, Apollo 9 successfully tested the lunar module in earth orbit, the first time it flew in space. Two months later, Apollo 10 was a complete dress rehearsal for the lunar landing, flying the lunar module in lunar orbit to within 50,000 feet of the surface directly over the Sea of Tranquility, the landing site for the next mission.

Set for July 1969, Apollo 11 was the culmination of America's moon program. When Neil Armstrong, Ed White's close friend, announced that "The *Eagle* has landed," and then set foot upon the moon on July 20, Kennedy's deadline had been met. But those early lunar pioneers became American heroes because of the sacrifice made by the crew of Apollo 1. As Gene Cernan said, when Neil Armstrong took the first step on the moon's surface, "he really stood on the shoulders of Gus and Ed and Roger." And when Apollo 12 flew in November, America had landed on the moon twice before the end of the decade and returned safely to earth.[4]

Unbeknownst to most Americans, if the fire had not occurred, and if Apollo 1 been a successful flight, history would have been forever altered. "One thing that would probably have been different if Gus had lived; the first guy to walk on the moon would have been Gus Grissom, not Neil Armstrong," wrote Deke Slayton. "Bob Gilruth and headquarters and I agreed on one thing, prior to the Apollo fire: if possible, one of the Mercury astronauts would have the first chance at being first on the moon. And at that time Gus was the one guy from the original seven who had the experience to press on through to the landing." Not to mention the fact that he was the only one who was still available— Shepard and Slayton were still grounded, Carpenter and Cooper were in the doghouse, Glenn had retired, and Schirra had arguments with Slayton and others over the flight schedule, so he had already said he was leaving after one Apollo flight, which would be one of the earlier test flights, not one of the later lunar missions. And some seemed to think Gus thought it was his rightful place in the space program. As Cernan wrote, "Everyone in the program knew that Gus firmly believed that when the first American stepped onto lunar soil, the name patch on his suit would read: GRISSOM."[5]

Before the end of the Apollo program, four additional lunar landings took place in 1971 and 1972. In all, twelve Americans walked on the moon's surface. But had it not been for the sacrifice of Gus Grissom, Ed White, and Roger Chaffee, there is little question that the nation would have failed to meet Kennedy's deadline. The Apollo spacecraft, as it was then constructed, would not have been able to pull off such a feat. And even if Apollo 1 had been successful, a tragedy was likely in a later mission, perhaps even in space. "It was terrible," noted astronaut Tom Stafford. "But we had to do so many things to that spacecraft—literally hundreds of things. If we had gone on using it as it was, which we probably would have done had it not been for the fire, we might have lost two or three crews up there." NASA engineer Jerry Goodman agreed. "We definitely would have had an accident on a later mission with that space-craft before the re-design."[6]

An accident in space would have been seen by the public as far worse and may have killed the program for good; several disasters surely would have. As space writer James Schefter has written, "If the Apollo 204 tragedy happened today, the Apollo program might have been cancelled. The space race could have ended, or gone eventually to the Russians." But it did not happen in 1967. "Times were different. The people and its government were stronger."[7]

There was growing opposition in Congress and among the general public, however, and some thought the tide against Apollo may be too great to overcome. Instead, what came out of it was a renewed resolve to see the program through and, with that resolve, a much better space-craft. "In the astronaut corps we marveled at the new Apollo spacecraft," Alan Shepard said, and gained confidence that it would be "safe for us to fly." Even Gus's mother, Cecile Grissom, understood that truthful, yet painful, fact. "My son had to give his life to make [the Apollo spacecraft] better," she said to a reporter after the second lunar landing.[8]

Another development was a much safer space program. Those who tried to criticize NASA for failing to learn from the mistakes of Apollo 1 couldn't have been more wrong. For one thing, in a dangerous, experi-mental program like spaceflight, which is, even today, still in relative infancy, accidents will always happen. There's simply no way to account for everything that might possibly go wrong. As Frank Borman has said, Apollo "was uncharted territory, and a very complicated machine, with a very difficult mission. And to expect that you weren't going to have problems was unrealistic." There was simply no way ever to promise that there would never be another accident. But after the fire, numerous safety protocols were added so that the spacecraft became as safe as it could be and performed flawlessly throughout the rest of the program, save the oxygen tank explosion on Apollo 13, yet even then the craft was able to withstand pressures it wasn't designed to withstand and perform tasks it was never designed to perform. The crew was able to return home safely with the incredible assistance of Mission Control. The Saturn rockets also had a stellar performance record and the Apollo spacecraft flew to the moon nine times, transported astronauts to Skylab three times, and docked with a Soviet Soyuz spacecraft in 1975 during the Apollo-Soyuz Test Project.[9]

As strange as it sounds, it was the fire that made all of it possible. That has been the consensus opinion for more than fifty years. In an interview thirty-two years after the tragedy, NASA deputy administrator Bob Seamans concurred. "I think you can say that as tragic as it was, if we had not had the accident then, we would not have gone to the Moon in the decade." Chris Kraft was a bit hesitant but had to say it. "I don't think we would have gotten to the moon in the sixties if we had not had the fire. That's a terrible thing to say, but I think it is true." The moon might have been achieved but would have taken longer and cost more. "We'd have flown, found problems, taken months to fix them, flown again, found more problems, taken more months.... We might not have landed on the moon until 1970 or '71." Gene Kranz, agreed. "The ultimate success of Apollo was made possible by the sacrifices of Grissom, White, and Chaffee. The accident profoundly affected everyone in the program. There was an unspoken promise on everyone's part to the three astronauts that their deaths would not be in vain."[10]

As one of NASA's leading flight controllers, Kranz called a meeting of all flight directors and flight controllers in the days after the fire and made what is one of NASA's most memorable speeches, laying down the law to every person involved in Mission Control, requiring them to write down a new mantra and never forget it:

> From this day forward, Flight Control will be known by two words: tough and competent. Tough means we are forever accountable for what we do or what we fail to do. We will never again compromise our responsibilities.... Competent means we will never take anything for granted.... Mission Control will be perfect. When you leave this meeting today

you will go to your office and the first thing you will do there is to write Tough and Competent on your blackboards. It will never be erased. Each day when you enter the room, these words will remind you of the price paid by Grissom, White, and Chaffee. These words are the price of admission to the ranks of Mission Control.[11]

Courtesy of the Library of Congress

When Apollo 11 landed on the moon, Congressman Gerald Ford, the minority leader of the United States House of Representatives, issued a press release asking for Americans to pray for the safe return of the astronauts. "But let us also say a prayer for Roger Chaffee, Gus Grissom, and Ed White," he said, "and pay tribute to them for the heroic contribution they made so that *Eagle* might land on the moon and return to planet earth."[12]

Politicians, journalists, astronauts and administrators all know it: had it not been for the fire, there would be no American flag on the moon. And because there is, we can thank Gus Grissom, Ed White, and Roger Chaffee.

NOTES

Introduction

1. Jules Verne, *From the Earth to the Moon* (New York: Scribner, Armstrong & Company, 1874), 25.
2. Harry Hurt III, *For All Mankind* (New York: Atlantic Monthly Press, 1988), 24, 72; Verne, *From the Earth to the Moon*, 32.
3. Verne, *From the Earth to the Moon*, 93.
4. Mike Wright, "The Disney-Von Braun Collaboration and Its Influence on Space Exploration," NASA, 1993, https://www.nasa.gov/centers/marshall/history/vonbraun/disney_article.html.
5. Erik Bergaust, *Murder on Pad 34: The Shocking Story of the Apollo Disaster—and Why It May Happen Again* (New York: G. P. Putnam's Sons, 1968), 212; George Leopold, *Calculated Risk: The Supersonic Life and Times of Gus Grissom* (West Lafayette, Indiana: Purdue University Press, 2016), 290.
6. Erlend A. Kennan and Edmund H. Harvey Jr., *Mission to the Moon: A Critical Examination of NASA and the Space Program* (New York: William Morrow & Co., Inc., 1969), 17.
7. Hugo Young, Bryan Silcock, and Peter Dunn, *Journey to Tranquility: The History of Man's Assault on the Moon* (London: Jonathan Cape, 1969).

Prologue
The Foundation: Early Race for Space

1. Sergei Khrushchev, "How Rockets Learned to Fly," Foreword to Von Hardesty and Gene Eisman, *Epic Rivalry: The Inside Story of the Soviet and American Space Race* (Washington, D.C.: *National Geographic*, 2007), viii.
2. Ibid.
3. "Khrushchev's Son Recalls Sputnik, Gagarin ascent in U.S.-Soviet Space Race," The World, July 16, 2019, https://www.pri.org/stories/2019-07-16/khrushchev-s-son-recalls-sputnik-gagarin-ascent-us-soviet-space-race.
4. Robert C. Seamans Jr., *Aiming at Targets: The Autobiography of Robert C. Seamans Jr.* (Washington, D.C.: NASA History Office, 1996) 61.
5. Evan Thomas, *Ike's Bluff: President Eisenhower's Secret Battle to Save the World* (New York: Back Bay Books, 2012), 253; Hurt, *For All Mankind*, 49.
6. Nikita Khrushchev, *Khrushchev Remembers* (Boston: Little, Brown and Company, 1970), 516–17.
7. Lyndon Baines Johnson, *The Vantage Point: Perspectives of the Presidency, 1963–1969* (New York: Holt, Rinehart and Winston, 1971), 272; Matthew Brzezinsky, *Red Moon Rising: Sputnik and the Rivalries That Ignited the Space Age* (New York: Bloomsbury Press, 2008), 175.

8. Johnson, *Vantage Point*, 271, 285.
9. Roger D. Launius, *Reaching for the Moon: A Short History of the Space Race* (New Haven, Connecticut: Yale University Press, 2019), 30.
10. Jay Barbree, *"Live from Cape Canaveral": Covering the Space Race, From Sputnik to Today* (Washington, D.C.: Smithsonian Books, 2007), 5.
11. Sherman Adams, *Firsthand Report: The Story of the Eisenhower Administration* (New York: Harper & Brothers, 1961), 415; Johnson, *Vantage Point*, 273.
12. Barbree, *"Live from Cape Canaveral,"* 2.
13. Deborah Cadbury, *The Space Race: The Untold Story of Two Rivals and Their Struggle for the Moon* (New York: Harper Perennial, 2005), 184.
14. Barbree, *"Live from Cape Canaveral,"* 5–6.
15. Young et al., *Journey to Tranquility*, 68; Johnson, *Vantage Point*, 273; Harlan Lebo, *100 Days: How Four Events in 1969 Shaped America* (Lanham, Maryland: Rowman & Littlefield, 2019), 42; Dwight D. Eisenhower, *The White House Years, Volume 2: Waging Peace, 1956–1961* (New York: Doubleday & Company, Inc., 1965), 205.
16. Douglas Brinkley, *American Moonshot: John F. Kennedy and the Great Space Race* (New York: HarperCollins, 2019), 13–14, 33, 35, 39, 43; Robert Stone and Alan Andres, *Chasing the Moon* (New York: Ballatine Books, 2019), 16–17.
17. Brinkley, *American Moonshot*, 21.
18. Young et al., *Journey to Tranquility*, 19; U.S. House of Representatives, "Review of the Space Program: Hearings before the House Committee on Science and Astronautics," Part I, January and February 1960, 136.
19. Brinkley, *American Moonshot*, 6–18.
20. Ibid., 11; Trevor Thomas, "The Moon Race: Godless Socialism vs. Faithful Americanism," American Thinker, July 22, 2019, https://www.americanthinker.com/articles/2019/07/the_moon_race_godless_socialism_vs_faithful_americanism.html.
21. Eisenhower Sputnik Conference Memo: https://www.archives.gov/education/lessons/sputnik-memo.
22. Roger D. Launius, *"Sputnik* and the Origins of the Space Age," NASA, https://history.nasa.gov/sputnik/sputorig.html.
23. Eisenhower to Harry Cecil Butcher, November 12, 1957, and Eisenhower to Arthur Bradford Eisenhower, November 8, 1957, Louis Galambos and Daun Van Ee, eds., *Papers of Dwight D. Eisenhower, Volume XVIII: The Presidency: Keeping the Peace* (Baltimore: Johns Hopkins University Press, 2001), 562, 551.
24. Irwin F. Gellman, *The President and the Apprentice: Eisenhower and Nixon, 1952–1961* (New Haven, Connecticut: Yale University Press, 2015), 486; Dwight D. Eisenhower, "The President's News Conference," October 9, 1957, https://www.presidency.ucsb.edu/documents/the-presidents-news-conference-308.
25. Barbree, *"Live from Cape Canaveral,"* 6.
26. Ibid., 8–9, 11; Johnson, *Vantage Point*, 273; Hurt, *For All Mankind*, 49–50.
27. Hurt, *For All Mankind*, 50; Donald K. Slayton, *Deke!* (New York: Forge, 1994), 90.

28. Hurt, *For All Mankind*, 50.
29. Barbree, *"Live from Cape Canaveral,"* 13.
30. Hurt, *For All Mankind*, 50.
31. Alan Wasser, "LBJ's Space Race: What We Didn't Know Then," The Space Review, June 20, 2005, https://www.thespacereview.com/article/396/1.
32. Barbree, *"Live from Cape Canaveral,"* 17; W. David Compton, "Where No Man Has Gone Before: A History of Apollo Lunar Exploration Missions," NASA Special Publication, 1989, https://history.nasa.gov/SP-4214/contents.html.
33. Hurt, *For All Mankind*, 51; Slayton, *Deke!*, 89; Johnson, *Vantage Point*, 267–77; Eisenhower Presidential News Conference, October 22, 1959; Robert L. Branyan and Lawrence H. Larsen, eds., *The Eisenhower Administration, 1953–1961: A Documentary History*, Volume 2 (New York: Random House, 1971), 1217.
34. Hurt, *For All Mankind*, 51; Compton, "Where No Man Has Gone Before"; T. Keith Glennan, *The Birth of NASA: The Diary of T. Keith Glennan* (NASA Special Publication, 1993), 13.
35. Compton, "Where No Man Has Gone Before."
36. Neil Armstrong, Oral History Interview, as quoted in Brinkley, *American Moonshot*, xix–xx.

I

The Beginning: Mercury

1. James Reston, as quoted in Brinkley, *American Moonshot*, xviii.
2. Betty Grissom and Henry Still, *Starfall* (New York: Thomas Y. Cromwell Company, 1974), 8–9; Virgil "Gus" Grissom, *Gemini: A Personal Account of Man's Venture into Space* (New York: The Macmillan Company, 1968), 17.
3. Grissom, *Gemini*, 17–18.
4. Ibid., 18; Grissom and Still, *Starfall*, 13.
5. Grissom, *Gemini*, 18.
6. Grissom and Still, *Starfall*, 22–23.
7. Ibid., 23.
8. Ibid., 24.
9. Ibid., 25.
10. Gus Grissom, "Proud to Help Out," in Scott Carpenter, et al., *We Seven* (New York: Simon and Schuster, 1962), 55; Grissom and Still, *Starfall*, 19, 26–27; Lily Koppel, *The Astronaut Wives Club* (New York: Grand Central Publishing, 2014), 6.
11. Mercury 7 Press Conference, https://history.nasa.gov/40thmerc7/documents.htm; Koppel, *The Astronaut Wives Club*, 6.
12. Grissom, *Gemini*, 18–19.
13. Grissom, "Proud to Help Out," *We Seven*, 55.
14. Grissom, *Gemini*, 19, 21; Grissom, "Proud to Help Out," *We Seven*, 56.
15. Frank Borman, *Countdown: An Autobiography* (New York: Silver Arrow Books, 1988), 87–88.
16. Barbree, *"Live from Cape Canaveral,"* 31.

17. Grissom, *Gemini*, 21–22; Grissom and Still, *Starfall*, 33.
18. Grissom, "Proud to Help Out," *We Seven*, 58, 55.
19. Grissom, *Gemini*, 22; Grissom and Still, *Starfall*, 55.
20. Grissom, "Proud to Help Out," *We Seven*, 57; Slayton, *Deke!*, 72; Wally Schirra, *Schirra's Space* (Annapolis, Maryland: BlueJacket Books, 1988) 60.
21. Mercury 7 Press Conference, https://history.nasa.gov/40thmerc7/documents.htm.
22. Schirra, *Schirra's Space*, 60; Grissom, "Proud to Help Out," *We Seven*, 57; Slayton, *Deke*, 73.
23. Alan Shepard and Deke Slayton, *Moonshot: The Inside Story of America's Apollo Moon Landings* (New York: Open Road, 2011), 56; Grissom and Still, *Starfall*, 60–61.
24. Barbree, "*Live from Cape Canaveral*," 32.
25. Slayton, *Deke!*, 74; Schirra, *Schirra's Space*, 63; Grissom and Still, *Starfall*, 51.
26. Slayton, *Deke!*, 79, 88; Ray E. Boomhower, *Gus Grissom: The Lost Astronaut* (Indianapolis: Indiana Historical Society Press, 2004), 100; John Boynton, Author Interview, July 17, 2020; Lowell Grissom, Author Interview, July 18, 2020.
27. Grissom, *Gemini*, xi; Slayton, *Deke!*, 79, 88; Leopold, *Calculated Risk*, 212.
28. Ed Buckbee with Wally Schirra, *The Real Space Cowboys* (Ontario: Apogee Books, 2005), 71, 73; Chris Kraft, *Flight: My Life in Mission Control* (New York: Dutton, 2005), 145; Jim Lovell and Jeffrey Kluger, *Apollo 13* (Boston: Houghton Mifflin Company, 2000), 26; Gordon Cooper, *Leap of Faith: An Astronaut's Journey into the Unknown* (New York: HarperCollins, 2000), 161.
29. Shepard and Slayton, *Moon Shot*, 129; Eugene Cernan, *The Last Man on the Moon: Astronaut Eugene Cernan and America's Race in Space* (New York: St. Martin's Press, 1999), 14; Walter Cunningham, *The All-American Boys: An Insider's Look at the U.S. Space Program* (New York: ibooks Inc., 2004), 10.
30. John T. Shaw, *Rising Star, Setting Sun: Dwight D. Eisenhower, John F. Kennedy, and the Presidential Transition That Changed America* (New York: Pegasus Books, 2018), 47.
31. Theodore C. Sorensen, Oral History Interview, March 26, 1964, JFK Library, https://www.jfklibrary.org/asset-viewer/archives/JFKOH/Sorensen%2C%20Theodore%20C/JFKOH-TCS-01/JFKOH-TCS-01; Brinkley, xx–xxi.
32. Shaw, *Rising Star, Setting Sun*, 23–25.
33. William E. Burrows, *This New Ocean: The Story of the First Space Age* (New York: The Modern Library, 1999), 305.
34. Ibid., 304.
35. Hurt, *For All Mankind*, 55; Bobby Baker, Senate Oral History, 2009–2010, *Politico*, https://www.politico.com/magazine/story/2013/11/sex-in-the-senate-bobby-baker-99530_Page3.html.
36. James E. Webb, Oral History, April 29, 1969, Lyndon B. Johnson Library, http://www.lbjlibrary.net/collections/oral-histories/webb-e.-james.html; Johnson, *The Vantage Point*, 278–79.
37. Slayton, *Deke!*, 95; Robert A. Caro, *The Years of Lyndon Johnson: The Passage of Power* (New York: Vintage Books, 2012), 173; Hurt, *For All Mankind*, 55;

Seamans, *Aiming at Targets*, 127; Borman, *Countdown*, 104; John Noble Wilford, *We Reach the Moon* (New York: Bantam Books, 1969), 56.

38. Barbree, *"Live from Cape Canaveral,"* 40.

39. James Schefter, *The Race: The Story of the Moon Race between Russia & America* (London: Arrow Books, 2000), 136.

40. Ibid., 135.

41. Shepard and Slayton, *Moon Shot*, 78.

42. Compton, "Where No Man Has Gone Before"; This is the general thesis of James Schefter's book, *The Race: The Story of the Moon Race between Russia & America*. Schefter was Time-Life's NASA correspondent from 1963–1973.

43. Cernan, *The Last Man on the Moon*, 129; Borman, *Countdown*, 97.

44. Slayton, *Deke!*, 91.

45. Eisenhower Presidential News Conference, October 22, 1959, in Branyan and Larsen, 1216.

46. John Lewis Gaddis, *The Cold War: A New History* (New York: The Penguin Press, 2005), 69; Grissom, *Gemini*, 12.

47. John F. Kennedy, Memorandum for Vice President, April 20, 1961, NASA Historical Reference Collection, NASA Headquarters, Washington, D.C., https://history.nasa.gov/Apollomon/apollo1.pdf; Hurt, *For All Mankind*, 51.

48. Seamans, *Aiming at Targets*, 82–83.

49. Ibid., 84–85.

50. Johnson, *The Vantage Point*, 280–81; Lyndon B. Johnson, "Evaluation of Space Program," Vice President, Memorandum for the President, NASA Historical Reference Collection, April 28, 1961, https://history.nasa.gov/Apollomon/apollo2.pdf.

51. Kraft, *Flight*, 142.

52. Ibid., 142–43.

53. Seamans, *Aiming at Targets*, 90–91; John F. Kennedy, "Special Message to Congress on Urgent National Needs," May 25, 1961, https://www.jfklibrary.org/archives/other-resources/john-f-kennedy-speeches/united-states-congress-special-message-19610525.

54. Shepard and Slayton, *Moonshot*, 122–23.

55. Compton, "Where No Man Has Gone Before"; Charles Murray and Catherine Bly Cox, *Apollo: The Behind-the-Scenes Story of One of Humankind's Greatest Achievements* (Connelly Springs, North Carolina: South Mountain Books, 2010), Kindle Edition; Kraft, *Flight*, 143; Mark Betancourt, "We Built the *Saturn V*," *Air and Space Magazine*, October 2017, https://www.airspacemag.com/space/we-built-saturn-v-180964759/.

56. Young et al., *Journey*, 197.

57. Gus Grissom, "The Trouble with *Liberty Bell*," *We Seven*, 206.

58. Ibid.

59. Kraft, *Flight*, 144.

60. Grissom, "The Trouble with *Liberty Bell*," *We Seven*, 205.

61. Ibid., 224.

62. Kraft, *Flight*, 145–46.

63. Buckbee, *The Real Space Cowboys*, 71; Grissom, "The Trouble with *Liberty Bell*," *We Seven*, 224–25.
64. Grissom, "The Trouble with *Liberty Bell*," *We Seven*, 225–26; Loyd Swenson, Jr., James M. Grimwod, and Charles C. Alexander, *This New Ocean: A History of Project Mercury* (NASA Special Publication, 1989), https://history.nasa.gov/SP-4201/toc.htm.
65. Kraft, *Flight*, 146.
66. Grissom, "The Trouble with *Liberty Bell*," *We Seven*, 225–27.
67. Slayton, *Deke!*, 100–1; Scott Carpenter, *For Spacious Skies: The Uncommon Journey of a Mercury Astronaut* (Orlando: Harcourt Books, 2002), 231.
68. Buckbee, *The Real Space Cowboys*, 71; Slayton, *Deke!*, 99.
69. Cooper, *Leap of Faith*, 32.
70. Grissom, "Trouble with *Liberty Bell*," *We Seven*, 227.
71. Swenson Jr. et al., *This New Ocean*; Buckbee, *The Real Space Cowboys*, 72–73.
72. Slayton, *Deke!*, 100; Buckbee, *The Real Space Cowboys*, 72–73.
73. Buckbee, *The Real Space Cowboys*, 72.
74. Kraft, *Flight*, 147.
75. Slayton, *Deke!*, 87.
76. John F. Kennedy, "Speech at Rice University," Houston, Texas, September 12, 1962, https://er.jsc.nasa.gov/seh/ricetalk.htm.
77. Schefter, *The Race*, 181.
78. "MA-9 Air-Ground Voice Communications," Mercury Project Summary (NASA SP-45), NASA, https://history.nasa.gov/SP-45/app.f.htm.
79. John F. Kennedy, "Remarks upon Presenting the NASA Distinguished Service Medal to Astronaut L. Gordon Cooper," May 21, 1963, *Public Papers of the Presidents: John F. Kennedy, 1963* (Washington, D.C.: U.S. Government Printing Office, 1964), 200.
80. Brinkley, *American Moonshot*, xi–xii.
81. Charles Fishman, *One Giant Leap: The Impossible Mission That Flew Us to the Moon* (New York: Simon and Schuster, 2019), 223; Ted Sorensen, *Counselor: A Life at the Edge of History* (New York: HarperCollins, 2008), 334; Sorensen, Oral History Interview, 1.
82. Fishman, *One Giant Leap*, 223.
83. John F. Kennedy, "Remarks at the Dedication of the Aerospace Medical Health Center," San Antonio, Texas, November 21, 1963, JFK Library, https://www.jfklibrary.org/archives/other-resources/john-f-kennedy-speeches/san-antonio-tx-19631121.

II

The Bridge: Gemini

1. Mary C. White, NASA Biography of Ed White, https://history.nasa.gov/Apollo204/zorn/white.htm.
2. Buzz Aldrin, *Men from Earth* (New York: Bantam Books, 1989), 31–32.
3. John Young, *Forever Young: A Life of Adventure in Air and Space* (Gainesville, Florida: University Press of Florida, 2012), 57.

4. Slayton, *Deke!*, 119.
5. Ibid., 119, 123; Schirra, *Schirra's Space*, 135; Cernan, *The Last Man on the Moon*, 73; Borman, *Countdown*, 96; Koppel, *The Astronaut Wives Club*, 108; Colin Burgess and Kate Doolan, *Fallen Astronauts: Heroes Who Died Reaching for the Moon* (Omaha, Nebraska: University of Nebraska Press, 2016), 178–79.
6. Cernan; *The Last Man on the Moon*, 130; Aldrin, *Men from Earth*, 117; Buckbee, *The Real Space Cowboys*, 73.
7. Grissom, *Gemini*, 14.
8. Michael Collins, *Liftoff: The Story of America's Adventure in Space* (New York: Grove Press, 1988), 67.
9. Buckabee, *The Real Space Cowboys*, 73; Grissom, *Gemini*, 73.
10. Grissom, *Gemini*, 8.
11. Ibid., 15–16; D. C. Agle, "Flying the Gusmobile," *Air and Space Magazine*, September 1998, https://www.airspacemag.com/flight-today/flying-the-gusmobile-218187/?all.
12. Buckbee, *The Real Space Cowboys*, 73.
13. Grissom, *Gemini*, 38; Buckabee, *The Real Space Cowboys*, 73; Agle, "Flying the Gusmobile"; Grissom and Still, *Starfall*, 142; Piers Bizony, *The Man Who Ran the Moon: James E. Webb, NASA, and the Secret History of Project Apollo* (New York: Thunder's Mouth Press, 2006), 105.
14. Grissom, *Gemini*, 11, 13.
15. Slayton, *Deke!*, 125, 132, 136–37.
16. Ibid., 137.
17. Ibid.; Koppel, *The Astronaut Wives Club*, 137; Young, *Forever Young*, 65; Grissom, *Gemini*, 74.
18. Grissom, *Gemini*, 2–3.
19. Ibid., 6.
20. Ibid., 98.
21. Slayton, *Deke!*, 148.
22. Neil Armstrong, Michael Collins, and Buzz Aldrin, *First on the Moon* (Boston: Little, Brown and Company, 1970), 46.
23. Slayton, *Deke!*, 148; Grissom, *Gemini*, 100.
24. Slayton, *Deke!*, 148; Grissom, *Gemini*, 102.
25. Grissom, *Gemini*, 107; Grissom and Still, *Starfall*, 151.
26. Grissom, *Gemini*, 108–9.
27. Ibid., 110–11.
28. Slayton, *Deke!*, 148–49; Grissom, *Gemini*, 112.
29. Grissom, *Gemini*, 111–12, 116; Agle, "Flying the Gusmobile."
30. Grissom, *Gemini*, 110.
31. Aldrin, *Men from Earth*, 127; Grissom and Still, *Starfall*, 153; Slayton, *Deke!*, 149; Young, *Forever Young*, 84.
32. Young, *Forever Young*, 82; Grissom, *Gemini*, 113.
33. Ibid.
34. Ibid., 117–19; Grissom and Still, *Starfall*, 157.
35. Slayton, *Deke!*, 101; Kraft, *Flight*, 144.

36. Slayton, *Deke!*, 101; "A History of the Committee on Science," House Committee on Science, https://web.archive.org/web/20060824092215/http://www.house.gov/science/committeeinfo/history/index.htm.

37. Caro, *Passage of Power*, 173.

38. Young et al., *Journey*, 112.

39. Norman Mailer, *Of a Fire on the Moon* (New York: Random House, 2014), 8; Hurt, *For All Mankind*, 54; Schefter, *The Race*, 150; Young et al., *Journey*, 162.

40. Wilford, *We Reach the Moon*, 63.

41. Robert Dallek, *Lyndon Johnson and His Times* (New York: Oxford, 1998), 22.

42. Wilford, *We Reach the Moon*, 63; Mailer, *Of a Fire on the Moon*, 8; Slayton, *Deke!*, 102; Mike Gray, *Angle of Attack: Harrison Storms and the Race to the Moon* (New York: W.W. Norton and Company, 1992), 104.

43. Borman, *Countdown*, 106; Koppel, *The Astronaut Wives Club*, 103–4; Cernan, *The Last Man on the Moon*, 344.

44. Slayton, *Deke!*, 150.

45. Aldrin, *Men from Earth*, 121.

46. Schefter, *The Race*, 204–6.

47. Ibid.

48. *Life*, June 18, 1965; Cernan, *The Last Man on the Moon*, 131.

49. Borman, *Countdown*, 123; James Salter, *Burning the Days* (New York: Random House, 1997), 285; Stone and Andres, *Chasing the Moon*, 153.

50. *Dallas Morning News*, October 20, 2018.

51. Cunningham, *The All-American Boys*, 11; Young, *Forever Young*, 59; Michael Collins, *Carrying the Fire: An Astronaut's Journeys* (New York: Farrar, Straus and Giroux, 2019), 144; Aldrin, *Men from Earth*, 31; Burgess and Doolan, *Fallen Astronauts*, 175, 188; Armstrong, NASA Oral History, September 19, 2001, https://historycollection.jsc.nasa.gov/JSCHistoryPortal/history/oral_histories/ArmstrongNA/armstrongna.htm; *San Antonio Light*, January 29, 1967.

52. Borman, *Countdown*, 123; 171, 294.

53. Ibid., 123.

54. Cernan, *The Last Man on the Moon*, 3, 73.

55. Aldrin, *Men from Earth*, 18; Mary C. White, NASA Biography of Ed White; Burgess and Doolan, *Fallen Astronauts*, 160–61, 179–80; "Airman Missing from the Vietnam War Accounted For (White)," Defense POW/MIA Accounting Agency, https://www.dpaa.mil/News-Stories/Recent-News-Stories/Article/1247773/airman-missing-from-the-vietnam-war-accounted-for-white/.

56. Mary C. White, NASA Biography of Ed White.

57. Ibid.

58. Burgess and Doolan, *Fallen Astronauts*, 162.

59. Ibid., 163.

60. Aldrin, *Men from Earth*, 31; Mary C. White, NASA Biography of Ed White; Burgess and Doolan, *Fallen Astronauts*, 179; Grissom, *Gemini*, 125.

61. Burgess and Doolan, *Fallen Astronauts*, 172–73; Koppel, *The Astronaut Wives Club*, x; Mary C. White, NASA Biography of Ed White.

62. Burgess and Doolan, *Fallen Astronauts*, 176.

63. This entire story, and the quotes used, came from Lovell, *Apollo 13*, 187, and Burgess and Doolan, *Fallen Astronauts*, 177–78.

64. *Life*, June 18, 1965.

65. Burgess and Doolan, *Fallen Astronauts*, 180.

66. Eugen Reichl, *Project Gemini* (Atglen, Pennsylvania: Schiffer Publishing, 2016), 65; Slayton, *Deke!*, 151; Cernan, *The Last Man on the Moon*, 121.

67. Burgess and Doolan, *Fallen Astronauts*, 181.

68. Ibid.

69. *Life*, June 18, 1965.

70. Slayton, *Deke!*, 151; Kraft, *Flight*, 220–21.

71. *Life*, June 18, 1965.

72. Burgess and Doolan, *Fallen Astronauts*, 185; Borman, *Countdown*, 120.

73. Hubert H. Humphrey, *The Education of a Public Man: My Life and Politics* (New York: Doubleday and Company, 1976), 414; Burgess and Doolan, *Fallen Astronauts*, 181.

74. Burgess and Doolan, *Fallen Astronauts*, 186; Lady Bird Johnson, *A White House Diary* (New York: Holt, Rinehart and Winston, 1970), 288–90.

75. Humphrey, *The Education of a Public Man*, 414–15; Slayton, *Deke!*, 158.

76. Burgess and Doolan, *Fallen Astronauts*, 186; Young, *Forever Young*, 83.

77. Andrew LePage, "Eight Days or Bust: The Mission of Gemini 5," Drew Ex Machina, August 21, 2015, https://www.drewexmachina.com/2015/08/21/eight-days-or-bust-the-mission-of-gemini-5/.

78. Shepard and Slayton, *Moon Shot*, 167.

79. Cernan, *The Last Man on the Moon*, 130.

80. Aldrin, *Men from Earth*, 145–46; Cernan, *The Last Man on the Moon*, 129–51.

81. Collins, *Carrying the Fire*, 122.

82. Aldrin, *Men from Earth*, 151–59.

83. Ibid.

III

The Goal: Apollo

1. C. Donald Chrysler and Don L. Chaffee, *On Course to the Stars: The Roger B. Chaffee Story* (Grand Rapids, Michigan: Kregel Publications, 1968), 73.

2. Ibid., 74–75.

3. Ibid., 15.

4. Burgess and Doolan, *Fallen Astronauts*, 189–90; Mary C. White, NASA Biography of Roger Chaffee.

5. Chrysler and Chaffee, *On Course to the Stars*, 53–55.

6. Ibid.

7. Ibid., 55–56.

8. Ibid.

9. Ibid., 61–62.

10. Mary C. White, NASA Biography of Roger Chaffee; Chrysler and Chaffee, *On Course to the Stars*, 63.

11. Mary C. White, NASA Biography of Roger Chaffee; Chrysler and Chaffee, *On Course to the Stars*, 63–65.
12. Chrysler and Chaffee, *On Course to the Stars*, 72.
13. Ibid., 80–82.
14. Mary C. White, NASA Biography of Roger Chaffee.
15. Ibid.
16. Chrysler and Chaffee, *On Course to the Stars*, 84, 86–87.
17. Slayton, *Deke!*, 132–34.
18. Cernan, *The Last Man on the Moon*, 3, 8–9, 134, 141.
19. Leopold, *Calculated Risk*, 222.
20. Chrysler and Chaffee, *On Course to the Stars*, 111; Grissom and Still, *Starfall*, 178; Leopold, *Calculated Risk*, 222.
21. Grissom and Still, *Starfall*, 177–78.
22. Chrysler and Chaffee, *On Course to the Stars*, 112.
23. Cernan, *The Last Man on the Moon*, 3, 9.
24. Compton, *Where No Man Has Gone Before*; Swenson et al., *This New Ocean*.
25. Ibid..
26. Wilford, *We Reach the Moon*, 58.
27. Hurt, *For All Mankind*, 54–55.
28. Slayton, *Deke!*, 124.
29. Eugen Reichl, *Project Apollo: The Early Years, 1960–1967* (Atglen, Pennsylvania: Schiffer Publishing, 2016), 4–5; Hurt, *For All Mankind*, 53; Charles D. Benson and William Barnaby Faherty, *Moonport: A History of Apollo Launch Facilities and Operations* (NASA Special Publications, 1978), https://history.nasa.gov/SP-4204/contents.html.
30. Hurt, *For All Mankind*, 17–18; W. David Woods, *How Apollo Flew to the Moon* (New York: Springer Books, 2008), 79; Lebo, *100 Days*, 82; Grissom, *Gemini*, 52; Stone and Andres, *Chasing the Moon*, 196; Mark Gray, "The Mighty Saturns: Saturn 1 and Saturn 1B," April 3, 2013, YouTube video, https://www.youtube.com/watch?v=_z1a1R7RUfM.
31. Hurt, *For All Mankind*, 17–18, 44, 61–62, 64–65; Cernan, *The Last Man on the Moon*, 186; Gray, *Angle of Attack*, 205; Mark Betancourt, "We Built the Saturn V," *Air and Space Magazine*, October 2017, https://www.airspacemag.com/space/we-built-saturn-v-180964759/; https://airandspace.si.edu/collection-objects/rocket-engine-turbo-pump-cutaway-f-1/nasm_A19751580000.
32. Hurt, *For All Mankind*, 17–18, 61–62, 64–65.
33. Ibid., 17; Lebo, *100 Days*, 92–93; "Vehicle Assembly Building," Ground Systems, NASA, https://www.nasa.gov/content/vehicle-assembly-building; "The Crawlers," Ground Systems, NASA, https://www.nasa.gov/content/the-crawlers; "Tread, Crawler-Transporter, Saturn V Rocket," Smithsonian Air and Space Museum, https://airandspace.si.edu/collection-objects/tread-crawler-transporter-saturn-v-rocket/nasm_A19730875001.
34. Lebo, *100 Days*, 88.
35. Slayton, *Deke!*, 146.
36. Ibid., 125–26; Wilford, *We Reach the Moon*, 41.
37. Wilford, *We Reach the Moon*, 42–48; Slayton, *Deke!*, 126–27; Gray, *Angle of Attack*, 121.

38. Apollo 204 Review Board Final Report, 4.1, NASA, https://history.nasa.gov/
Apollo204/content.html; Charles Benson and William Faherty, *Moonport: A
History of Apollo Launch Facilities and Operations* (NASA Special
Publications, 1978); Slayton, *Deke!*, 193; Young, *Forever Young*, 111; Hurt, *For
All Mankind*, 34; Aldrin, *Men from Earth*, 163.
39. Schirra, *Schirra's Space*, 180.
40. Donn Eisele, *Apollo Pilot: The Memoir of Astronaut Donn Eisele* (Lincoln,
Nebraska: University Press of Nebraska, 2017), 32; Schirra, *Schirra's Space*, 181;
Slayton, *Deke!*, 168, 182–83.
41. Apollo 204 Review Board Final Report, 4.1.
42. Al Worden, *Falling to Earth: An Apollo 15 Astronaut's Journey* (Washington,
D.C.: Smithsonian Books, 2011), 84.
43. Thomas P. Stafford, *We Have Capture: Tom Stafford and the Space Race*
(Washington, D.C.: Smithsonian Books, 2002), Kindle Edition.
44. Frank Borman, NASA Oral History, April 13, 1999, https://historycollection.
jsc.nasa.gov/JSCHistoryPortal/history/oral_histories/BormanF/
Bormanff_4-13-99.htm.
45. Barbree, *"Live From Cape Canaveral,"* 123; Shepard and Slayton, *Moon Shot*,
178; Jeffrey Kluger, *Disaster Strikes! The Most Dangerous Space Missions of
All Time* (New York: Philomel Books, 2019), 57.
46. Young, *Forever Young*, 112; Jim Lovell, NASA Oral History, May 25, 1999,
https://historycollection.jsc.nasa.gov/JSCHistoryPortal/history/oral_histories/
LovellJA/LovellJA_5-25-99.htm.
47. Kluger, *Apollo 13*, 58–59.
48. Ibid., 57; Leopold, *Calculated Risk*, 211, 237, 252.
49. Barbree, *"Live From Cape Canaveral,"* 124–25. "Mac" referred to the
McDonnell Aircraft Company, which had built but the spacecraft for Mercury
and Gemini.
50. Grissom and Still, *Starfall*, 179.
51. Borman, NASA Oral History.
52. Barbree, *"Live From Cape Canaveral,"* 125; Leopold, *Calculated Risk*, 228.
53. Barbree, *"Live From Cape Canaveral,"* 126.
54. Stafford, *We Have Capture*.
55. Murray and Cox, *Apollo*; Gray, *Angle of Attack*, 181.
56. Benson and Faherty, *Moonport*; Apollo 204 Review Board Final Report, 4.1;
Schirra, *Schirra's Space*, 241.
57. Benson and Faherty, *Moonport*; Apollo 204 Review Board Final Report, 4.1, 4.2.
58. Apollo 204 Review Board Final Report, 4.1, 4.2.
59. Ibid.
60. Grissom and Still, *Starfall*, 178.
61. United States Senate, "Hearings before the Committee on Aeronautics and
Space Sciences," 90th Congress, First Session, January–February 1967, 208;
Apollo 204 Review Board Final Report, 4.2.
62. Shepard and Slayton, *Moon Shot*, 178.
63. Wilford, *We Reach the Moon*, 124; Mark E. Byrnes, *Politics and Space: Image
Making by NASA* (Westport, Connecticut: Praeger Publishers, 1994), 85.

64. Mailer, *Of a Fire on the Moon*, 373–74; Wilford, *We Reach the Moon*, 127; Alexis C. Madrigal, "Moondoggle: The Forgotten Opposition to the Apollo Program," *The Atlantic*, September 12, 2012, https://www.theatlantic.com/technology/archive/2012/09/moondoggle-the-forgotten-opposition-to-the-apollo-program/262254/.

65. Wilford, *We Reach the Moon*, 124; W. W. Rostow, *The Diffusion of Power: An Essay in Recent History* (New York: The Macmillan Company, 1972), 181.

66. *Washington Post*, July 20, 1989; James Jeffrey, "The Dark Side of the Moon Mission," Progressive.org, July 18, 2019, https://progressive.org/dispatches/dark-side-of-moon-mission-jeffrey-190718/; Burrows, *This New Ocean*, 423. The last quote by Vonnegut was used in the film, *First Man*.

67. Wilford, *We Reach the Moon*, 125.

68. Alexis C. Madrigal, "Gil Scott-Heron's Poem, 'Whitey on the Moon,'" *The Atlantic*, May 28, 2011, https://www.theatlantic.com/technology/archive/2011/05/gil-scott-herons-poem-whitey-on-the-moon/239622/.

69. "Ballistic Missile and Space Program – Draft – Fact Sheet," November 19, 1959, Branyan and Larsen, *The Eisenhower Administration*, 1219; Hurt, *For All Mankind*, 80; Roger D. Launius and Howard D. McCurly, eds., *Spaceflight and the Myth of Presidential Leadership* (Champaign, Illinois: University of Illinois Press, 1997), 226.

70. Dwayne Day, "When Senator Walter Mondale Went to the Moon," *Space Review*, March 16, 2020, https://www.thespacereview.com/article/3901/1; Congressional Record, Volume 109, Part 17, 22368.

71. Gray, *Angle of Attack*, 221.

72. Kenneth P. O'Donnell and Dave F. Powers, *Johnny, We Hardly Knew Ye: Memoirs of John F. Kennedy* (Boston: Little, Brown and Co., 1970), 410.

73. Ibid., 224–38.

74. W. D. Kay, *Defining NASA: The Historical Debate over the Agency's Mission*, 92; NASA, *NASA Historical Data Book, 1958–1968*, Volume I, 1976; Day, "When Senator Mondale Went to the Moon."

75. Wilford, *We Reach the Moon*, 125–26.

76. Byrnes, *Politics and Space*, 85.

77. Grissom, *Gemini*, 7.

78. Stone and Andres, *Chasing the Moon*, 177.

79. Slayton, *Deke!*, 165.

80. Gene Kranz, *Failure Is Not an Option: Mission Control from Mercury to Apollo 13 and Beyond* (New York: Simon and Schuster, 2000), 193; Borman, *Countdown*, 173.

IV

The Fire: "We're Burning Up!"

1. Armstrong et al., *First on the Moon*, 48; Lola Morrow, quoted in *Moonshot* documentary, at 1:44:02 to 1:44:25.

2. Slayton, *Deke!*, 193; Schirra, *Schirra's Space*, 184.

3. Leopold, *Calculated Risk*, 198. An often-used quote from Gus is: "If we die, we want people to accept it. We're in a risky business, and we hope that if anything happens to us it will not delay the program. The conquest of space is worth the risk of life." But there is no evidence that Gus ever said this. It has been alleged that Gus said something similar to Jules Bergman. See Leopold, *Calculated Risk*.
4. Koppel, *The Astronaut Wives Club*, 7.
5. Leopold, *Calculated Risk*, 223, 226; Burgess and Doolan, *Fallen Astronauts*, 199.
6. Leopold, *Calculated Risk*, 191; Cunningham, *The All-American Boys*, 6.
7. Grissom, *Gemini*, 8–9.
8. Cunningham, *The All-American Boys*, 8.
9. Slayton, *Deke!*, 188; Shepard and Slayton, *Moon Shot*, 179, 181.
10. Cooper, *Leap of Faith*, 161.
11. Leopold, *Calculated Risk*, 235–36; Burrows, *This New Ocean*, 409.
12. Shepard and Slayton, *Moon Shot*, 179.
13. James Donovan, *Shoot for the Moon: The Space Race and the Extraordinary Voyage of Apollo 11* (New York: Little, Brown and Company, 2019), 210–11; Eisele, *Apollo Pilot*, 33; Stafford, *We Have Capture*; Shepard and Slayton, *Moon Shot*, 184.
14. Shepard and Slayton, *Moon Shot*, 180; Schirra, *Schirra's Space*, 183.
15. Slayton, *Deke!*, 188–89.
16. Armstrong et al., *First on the Moon*, 47–48.
17. Kranz, *Failure Is Not an Option*, 195; Courtney G. Brooks, James M. Grimwood, and Loyd S. Swenson Jr., *Chariots for Apollo: A History of Manned Lunar Spacecraft* (NASA History Series, 1979), https://www.hq.nasa.gov/office/pao/History/SP-4205/cover.html.
18. Shepard and Slayton, *Moon Shot*, 183; Benson and Faherty, *Moonport*.
19. Shepard and Slayton, *Moon Shot*, 185; Slayton, *Deke!*, 189.
20. Nancy Atkinson, *Eight Years to the Moon: The History of the Apollo Missions* (Salem, Massachusetts: Page Street Publishing, 2019), 138–39.
21. Shepard and Slayton, *Moon Shot*, 185; Slayton, *Deke!*, 189.
22. Audio can be heard here: https://www.youtube.com/watch?v=274lQSbpkRg.
23. Shepard and Slayton, *Moon Shot*, 185.
24. Kluger, *Apollo 13*, 64–65.
25. Andrew Chaikin, "Apollo's Worst Day," *Air and Space Magazine*, November 2016, https://www.airspacemag.com/history-of-flight/apollo-fire-50-years-180960972/.
26. Slayton, *Deke!*, 189; Barbree, *"Live From Cape Canaveral,"* 129.
27. Barbree, *"Live From Cape Canaveral,"* 129; Chaikin, "Apollo's Worst Day"; Lovell, *Apollo 13*, 17.
28. Benson and Faherty, *Moonport*.
29. Shepard and Slayton, *Moon Shot*, 188–89; Slayton, *Deke!*, 189; Apollo 204 Review Board Final Report, 4.6; Benson and Faherty, *Moonport*.
30. Slayton, *Deke!*, 189; Gray, *Angle of Attack*, 229.
31. Margi Murphy, "'I Heard Them Scream': *Apollo 1* Crew's Last Moments Revealed by NASA Flight Director Chris Kraft," *The Sun*, January 24, 2017, https://www.thesun.co.uk/news/2692032/apollo-1-crews-last-moments-revealed-by-nasa-flight-director-kris-kraft-who-listened-in-from-mission-control/.

32. Atkinson, *Eight Years to the Moon*, 139.
33. Murphy, "'I Heard Them Scream'"; Cunningham, *The All-American Boys*, 9.
34. Cunningham, *The All-American Boys*, 8.
35. Young, *Forever Young*, 113.
36. Shepard and Slayton, *Moon Shot*, 191; Donovan, *Shoot for the Moon*, 215.
37. Slayton, *Deke!*, 189.
38. Chrysler and Chaffee, *On Course to the Stars*, 127; Leopold, *Calculated Risk*, 261.
39. Slayton, *Deke!*, 190; Armstrong et al., *First on the Moon*, 48–49; Shepard and Slayton, *Moon Shot*, 191; Schirra, *Schirra's Space*, 183.
40. Worden, *Falling to Earth*, 85; Young, *Forever Young*, 112.
41. Stone and Andres, *Chasing the Moon*, 178; Johnson, *Vantage Point*, 270
42. Stone and Andres, *Chasing the Moon*, 178; Slayton, *Deke!*, 190; Aldrin, *Men from Earth*, 165–66.
43. Stone and Andres, *Chasing the Moon*, 178–79; Johnson, *The Vantage Point*, 270–71; Johnson, *A White House Diary*, 482.
44. Shepard and Slayton, *Moon Shot*, 196.
45. Seamans, *Aiming at Targets*, 137.
46. Lovell, *Apollo 13*, 22–23.
47. Cooper, *Leap of Faith*, 163.
48. Lovell, *Apollo 13*, 24.
49. Shepard and Slayton, *Moon Shot*, 191.
50. Charlie and Dotty Duke, *Moonwalker* (Nashville, Tennessee: Thomas Nelson Inc., 1990).
51. Kranz, *Failure Is Not an Option*, 197.
52. Ibid., 199; Kraft, *Flight*, 271.
53. Slayton, *Deke!*, 190; Armstrong et al., *First on the Moon*, 49; Seamans, *Aiming at Targets*, 138.
54. *Life*, January 26, 1968; Koppel, *The Astronaut Wives Club*, 123.
55. Koppel, *The Astronaut Wives Club*, 162–63.
56. Ibid., 164; Grissom and Still, *Starfall*, 189.
57. Leopold, *Calculated Risk*, 266–67; Cunningham, *The All-American Boys*, 10.
58. *Life*, January 26, 1968; Cunningham, *The All-American Boys*, 11; Koppel, *The Astronaut Wives Club*, 167; Borman, *Countdown*, 294–96; *Dallas Morning News*, October 20, 2018.
59. Koppel, *The Astronaut Wives Club*, 164–65; *Life*, January 26, 1968; Cunningham, *The All-American Boys*, 9.
60. *New York Times*, January 28, 1967.
61. Ibid.
62. Ibid.
63. Ibid.
64. Shepard and Slayton, *Moon Shot*, 197.
65. Stone and Andres, *Chasing the Moon*, 183.
66. *Newsweek*, February 13, 1967; Benson and Faherty, *Moonport*; *New York Times*, April 11, 1967.
67. *New York Times*, January 30, 1967; *The Nation*, February 13, 1967.

68. Borman, *Countdown*, 167; Cunningham, *The All-American Boys*, 10; *Aviation Week*, February 6, 1967.

69. Chrysler and Chaffee, *On Course to the Stars*, 127.

70. Ibid.; Byrnes, *Politics and Space*, 85; Lovell, *Apollo 13*, ix.

71. *San Antonio Light*, January 29, 1967.

72. Schirra, *Schirra's Space*, 183; Koppel, *The Astronaut Wives Club*, 166.

73. Borman, *Countdown*, 169–70.

74. Lovell, *Apollo 13*, 26; Lowell Grissom, Interview with the Author, July 18, 2020.

75. Donovan, 222; Slayton, *Deke!*, 191; Lovell, *Apollo 13*, 25; Aldrin, *Men from Earth*, 169; Lady Bird Johnson, *A White House Diary*, 484.

76. Slayton, *Deke!*, 191–92; Shepard and Slayton, *Moon Shot*, 196.

77. Cunningham, *The All-American Boys*, 4; Armstrong et al., *First on the Moon*, 49.

78. Worden, *Falling to Earth*, 86.

V
The Investigations: "Stop the Witch Hunt"

1. Slayton, *Deke!*, 137; Young, *Forever Young*, 114; Cernan, *The Last Man on the Moon*, 74.

2. Murray and Cox, *Apollo*.

3. Johnson, *The Vantage Point*, 271; Hurt, *For All Mankind*, 15.

4. Shepard and Slayton, *Moon Shot*, 199.

5. Ibid.

6. Slayton, *Deke!*, 190; Young, *Forever Young*, 111, 113, 116; Chaikin, "Apollo's Worst Day"; Armstrong et al., *First on the Moon*, 47.

7. Shepard and Slayton, *Moon Shot*, 199.

8. Cunningham, *The All-American Boys*, 12.

9. Stone and Andres, *Chasing the Moon*, 186–87.

10. Young, *Forever Young*, 114; Seamans, *Aiming at Targets*, 138; Brooks et al., *Chariots for Apollo*.

11. Borman, *Countdown*, 174; Seamans, *Aiming at Targets*, 138; Stafford, *We Have Capture*.

12. Borman, *Countdown*, 169–70.

13. Ibid., 172–73.

14. Ibid., 173–74.

15. Cunningham, *The All-American Boys*, 12–13; Eisele, *Apollo Pilot*, 33.

16. Eisele, *Apollo Pilot*, 33.

17. Benson and Faherty, *Moonport*; Slayton, *Deke!*, 192; Murray and Cox, *Apollo*.

18. Gray, *Angle of Attack*, 248; Borman, *Countdown*, 174; Young, *Forever Young*, 114.

19. Apollo 204 Review Board Final Report, 5.12.

20. Apollo 204 Review Board Final Report, 5.9.

21. Young, *Forever Young*, 111, 115; Apollo 204 Review Board Final Report, 5.11.

22. Apollo 204 Review Board Final Report, 5.11; Schirra, *Schirra's Space*, 182; Lovell, *Apollo 13*, 13; Young, *Forever Young*, 111–12, 115.

23. Borman, *Countdown*, 174; Murray and Cox, *Apollo*; Apollo 204 Review Board Final Report, 5.10; Chaikin, "Apollo's Worst Day"; *New York Times*, September 13, 1971.
24. Borman, *Countdown*, 175, 178.
25. Young, *Forever Young*, 113–14.
26. Apollo 204 Review Board Final Report, 5.10.
27. Borman Testimony, Senate Hearings, 208; Apollo 204 Review Board Final Report, 5.10.
28. Young, *Forever Young*, 114.
29. Young, *Forever Young*, 65, 114; Leopold, *Calculated Risk*, 227.
30. Borman, *Countdown*, 174–75.
31. Cunningham, *The All-American Boys*, 13.
32. Benson and Faherty, *Moonport*.
33. Ibid.
34. Ibid.
35. Ibid.; Gary W. Johnson, Email to the Author, July 18, 2020.
36. Brooks et al., *Chariots for Apollo*; Chaikin, "Apollo's Worst Day."
37. Korb, *Memories of the Apollo and Space Shuttle Programs.*
38. Teitel, "Why Apollo Had a Flammable Pure Oxygen Environment."
39. *Amarillo Daily News*, February 6, 1967.
40. Worden, *Falling to Earth*, 88; Leopold, *Calculated Risk*, 231; Young, et al., *Journey*, 229.
41. Glynn Lunney, NASA Oral History, March 9, 1998, https://historycollection.jsc. nasa.gov/JSCHistoryPortal/history/oral_histories/LunneyGS/LunneyGS_3-9-98. htm; Lovell, NASA Oral History; Young, *Forever Young*, 112.
42. Schirra, *Schirra's Space*, 185.
43. Worden, *Falling to Earth*, 87; Eisele, *Apollo Pilot*, 34; Leopold, *Calculated Risk*, 229.
44. Eisele, *Apollo Pilot*, 34; Apollo 204 Review Board Final Report, 6.1; Senate Hearing, February 7, 1967, 97–98.
45. Murray and Cox, *Apollo*.
46. Slayton, *Deke!*, 195; Bergaust, *Murder on Pad 34*, 141.
47. Slayton, *Deke!*, 195.
48. The following portion of congressional testimony was also taken from Bergaust.
49. House of Representatives, Investigation into Apollo 204 Accident, 90th Congress, First Session, Volume I, 408.
50. Borman, *Countdown*, 175–76; Slayton, *Deke!*, 192; Boynton Interview, July 17, 2020.
51. Seamans, *Aiming at Targets*, 139; Bizony, *The Man Who Ran the Moon*, 144.
52. Borman, *Countdown*, 175–76; Murray and Cox, *Apollo*; Seamans, *Aiming at Targets*, 139–40.
53. Borman, *Countdown*, 178.
54. Slayton, *Deke!*, 192–93; Seamans, *Aiming at Targets*, 140.
55. Kraft, *Flight*, 275.
56. Boomhower, *Gus Grissom*, 319.
57. Seamans, *Aiming at Targets*, 140.

58. Ibid.
59. Wilford, *We Reach the Moon*, 104.
60. Borman, *Countdown*, 186; Wilford, *We Reach the Moon*, 64.
61. "The Phillips Report," NASA Historical Reference Collection, 1965, https://history.nasa.gov/Apollo204/phillip1.html; Wilford, *We Reach the Moon*, 107; Gray, *Angle of Attack*, 120.
62. "The Phillips Report."
63. Bizony, *The Man Who Ran the Moon*, 123–24; Dwayne A. Day, "Capsule on Fire: An Interview with Robert Seamans about the *Apollo 1* Accident," *The Space Review*, March 23, 2020, https://www.thespacereview.com/article/3904/1.
64. Senate Hearing, February 7, Part 1, 126–27.
65. Bizony, *The Man Who Ran the Moon*, 124.
66. Day, "When Senator Mondale Went to the Moon."
67. Seamans, *Aiming at Targets*, 141.
68. Ibid.
69. Ibid.; Phillips Report; Young et al., *Journey*, 244–45.
70. Bizony, *The Man Who Ran the Moon*, 124.
71. Kennan and Harvey, *Mission to the Moon*, 115–16; Benson and Faherty, *Moonport*; Brooks et al., *Chariots for Apollo*; *New York Times*, April 30, 1967; Bizony, *The Man Who Ran the Moon*, 136–39.
72. Senate Hearing, 131.
73. Day, "When Senator Mondale went to the Moon."
74. Seamans, *Aiming at Targets*, 141–42.
75. Day, "Capsule on Fire"; Bizony, *The Man Who Ran the Moon*, 127.
76. Bizony, *The Man Who Ran the Moon*, 127.
77. *Washington Star*, May 21, 1967, as quoted in Bizony, *The Man Who Ran the Moon*, 147–48.
78. Day, "When Senator Mondale went to the Moon."
79. *New York Times*, April 27–29, 1967; *Minneapolis Tribune*, June 4, 1967.
80. Wilford, *We Reach the Moon*, 104; Slayton, *Deke!*, 194.
81. Seamans, *Aiming at Targets*, 142.
82. U.S. Senate, Report of the Committee on Aeronautical and Space Sciences, Apollo 204 Accident, January 30, 1968, 11.
83. Andrew Chaikin, *A Man on the Moon: The Voyages of the Apollo Astronauts* (London: Penguin, 1994), 26.
84. Grissom and Still, *Starfall*, 199; Congressional Record, April 10, 1967, 3786–87; Benson and Faherty, *Moonport*; Murray and Cox, *Apollo*.
85. Chaikin, "Apollo's Worst Day"; Slayton, *Deke!*, 194–95.
86. Borman, *Countdown*, 178; Aldrin, *Men from Earth*, 167; Gray, *Angle of Attack*, 245.
87. Bergaust, *Murder on Pad 34*, 151–53.
88. *New York Times*, May 11, 1967.
89. Wilford, *We Reach the Moon*, 107; Schirra, *Schirra's Space*, 187; House Hearing, Volume I, 544; Bizony, *The Man Who Ran the Moon*, 144.
90. Borman, *Countdown*, 178–79.

91. Borman, *Countdown*, 180; Robert Kurson, *Rocket Men: The Daring Odyssey of Apollo 8 and the Astronauts Who Made Man's First Journey to the Moon* (New York: Random House, 2019), 11; Stone and Andres, *Chasing the Moon*, 193.
92. Schefter, *The Race*, 253.
93. Young et al., *Journey*, 246; Borman, *Countdown*, 180.
94. Apollo 204 Accident, Senate Report, January 30, 1968, iii.
95. Ibid., 15–16.
96. Stafford, *We Have Capture*; Shepard and Slayton, *Moon Shot*, 201.
97. Borman, *Countdown*, 181–83; Stafford, *We Have Capture*; Aldin, *Men from Earth*, 168.
98. Korb, *Memories of the Apollo and Space Shuttle Programs*.
99. Schirra, *Schirra's Space*, 178; Borman, *Countdown*, 182; Seamans, *Aiming at Targets*, 145.
100. Slayton, *Deke!*, 200; Author Interview with Jerry Goodman, July 17, 2020.
101. Goodman Interview; Young, *Forever Young*, 115; Borman, *Countdown*, 184–86; Cunningham, *The All-American Boys*, 17; Shepard and Slayton, *Moon Shot*, 203; Wilford, *We Reach the Moon*, 109.
102. Burgess and Doolan, *Fallen Astronauts*, 213.
103. Hurt, *For All Mankind*, 29.
104. Leopold, *Calculated Risk*, 206–7.
105. Ibid; Hurt, *For All Mankind*, 31.
106. Young, *Forever Young*, 116.
107. Eisele, *Apollo Pilot*, 33–34.
108. Ibid., 34.
109. Leopold, *Calculated Risk*, 294; Lowell, Grissom Interview.
110. Johnson, *The Vantage Point*, 284.

VI

The Politics: A Webb of Intrigue

1. Joan Mellen, *Faustian Bargains: Lyndon Johnson and Mac Wallace in the Robber Baron Culture of Texas* (New York: Bloomsbury, 2016), 112–13; Bobby Baker, *Wheeling and Dealing: Confessions of a Capitol Hill Operator* (New York: W. W. Norton, 1978), 34–35.
2. Baker, *Wheeling and Dealing*, 34–35; Phillip F. Nelson, *LBJ: The Mastermind of the JFK Assassination* (New York: Skyhorse Publishing, 2011), 40; Baker, Senate Oral History, 2009–2010.
3. Mellen, *Faustian Bargains*, 112–13; Bizony, *The Man Who Ran the Moon*, 151.
4. Mailer, *Of a Fire on the Moon*, 170.
5. Brooks et al., *Chariots for Apollo*; Schirra, *Schirra's Space*, 178; Seamans, *Aiming at Targets*, 96; Gray, *Aiming at Targets*, 113.
6. Brooks et al., *Chariots for Apollo*; Gray, *Angle of Attack*, 104, 115.
7. Murray and Cox, *Apollo*.
8. Gray, *Angle of Attack*, 115.
9. Murray and Cox, *Apollo*; *Minneapolis Tribune*, June 4, 1967.
10. Ibid.

11. Schirra, *Schirra's Space*, 178–79; Young, *Forever Young*, 116.
12. Mailer, *Of a Fire on the Moon*, 171; *New York Times*, April 17, 1967, as quoted in Wilford, *We Reach the Moon*, 108; Vice President's Daily Diary Entry, August 21, 1963, LBJ Library, https://www.discoverlbj.org/item/vpdd-19630821, and July 17, 1963, https://www.discoverlbj.org/item/vpdd-19630717.
13. Baker, Senate Oral History, 2009–2010.
14. Baker, *Wheeling and Dealing*, 170.
15. Ibid., 102; *New York Times*, January 31, 1964; Young et al., *Journey*, 150.
16. Baker, *Wheeling and Dealing*, 103–4; Gray, *Angle of Attack*, 118.
17. Baker, *Wheeling and Dealing*, 103–4; Webb, LBJ Oral History.
18. Wilford, *We Reach the Moon*, 108–9.
19. Clark Mollenhoff, "Webb: Man on a Hot Seat in Apollo Probe – The Moon Project's Tarnished Image," *Des Moines Register*, April 19, 1967, as quoted in Congressional Record, 90th Congress, Volume 113, 12388; Congressional Record, Volume 114, 28816; *Minneapolis Star Tribune*, June 4, 1967.
20. Seamans, *Aiming at Targets*, 102; Gray, *Angle of Attack*, 103.
21. Baker, *Wheeling and Dealing*, 103–4.
22. *New York Times*, January 31, 1964; Mollenhoff, *Des Moines Register*, as quoted in Congressional Record. In 2015, Texas congressman Lamar Smith made a statement about the waterway's involvement with the space program: "Tulsa's involvement in space manufacturing goes back many decades. The need to transport large aerospace components even figured into the completion of the McClellan-Kerr Arkansas River Navigation System," https://tulsaworld.com/news/local/government-and-politics/jim-bridenstine-committee-chairman-lamar-smith-talk-space-environmental-regulation/article_084e479c-07e5-5e7e-81e7-80ca018c5fd0.html.
23. Baker, *Wheeling and Dealing*, 169–70; Gray, *Angle of Attack*, 117–18.
24. Mellen, *Faustian Bargains*, 157–58; Bizony, *The Man Who Ran the Moon*, 153; Mailer, *Of a Fire on the Moon*, 171. For more on Kerr's relationship to Baker, see telephone conversation #1384 between Lyndon Johnson and Attorney General Ramsey Clark on January 20, 1967, LBJ Library, https://www.discoverlbj.org/item/tel-11384.
25. Mellen, *Faustian Bargains*, 157–58; Gray, *Angle of Attack*, 118.
26. Mildred Stegall and Bobby Baker, Telephone Conversation #8685, August 31, 1965, LBJ Library, https://www.discoverlbj.org/item/tel-08685; Mollenhoff, *Des Moines Register* as quoted in Congressional Record.
27. Clark Mollenhoff, "Apollo Fire Started Barrage of Questions about Handling," *Minneapolis Star Tribune*, June 4, 1967.
28. Baker, *Wheeling and Dealing*, 170–71.
29. Webb, LBJ Oral History.
30. Mollenhoff, "Apollo Fire Started Barrage of Questions about Handling."
31. Ibid.; Mollenhoff, *Des Moines Register* as quoted in Congressional Record.
32. Young, *Forever Young*, 116.

Afterword

The Moon: "The *Eagle* Has Landed"

1. Seamans, *Aiming at Targets*, 3.
2. John MacArthur, *Twelve Unlikely Heroes* (Nashville: Nelson Books, 2014), x.
3. Burrows, *This New Ocean*, 409.
4. Hal Boedeker, "Saga of Space Deserves More Time," *Orlando Sentinel*, November 21, 1996, https://www.orlandosentinel.com/news/os-xpm-1996-11-21-9611200057-story.html.
5. Slayton, *Deke!*, 191; Cernan, *The Last Man on the Moon*, 3.
6. Armstrong et al., *First on the Moon*, 49; Goodman Interview.
7. Schefter, *The Race*, 248.
8. Shepard and Slayton, *Moon Shot*, 201; Hurt, *For All Mankind*, 83; Leopold, *Calculated Risk*, 204.
9. Borman, NASA Oral History.
10. Kranz, *Failure Is Not an Option*, 205; Schefter, *The Race*, 249; Day, Seamans Interview, *Space Review*; Leopold, 275.
11. Kranz, *Failure Is Not an Option*, 204.
12. Gerald Ford, Press Releases, January 7, 1969, Ford Congressional Papers, Gerald R. Ford Presidential Library and Museum, https://www.fordlibrarymuseum.gov/library/document/0054/12130687.pdf.

INDEX